Editors

Michał Kunicki
Opole University of Technology
Opole
Poland

Jan Fulneček
VSB Technical University of Ostrava
Ostrava
Czech Republic

Pawel Rozga
Lodz University of Technology
Lodz
Poland

Editorial Office
MDPI AG
Grosspeteranlage 5
4052 Basel, Switzerland

This is a reprint of articles from the Special Issue published online in the open access journal *Sensors* (ISSN 1424-8220) (available at: https://www.mdpi.com/journal/sensors/special_issues/050PA74ZGH).

For citation purposes, cite each article independently as indicated on the article page online and as indicated below:

Lastname, A.A.; Lastname, B.B. Article Title. *Journal Name* **Year**, *Volume Number*, Page Range.

ISBN 978-3-7258-2243-0 (Hbk)
ISBN 978-3-7258-2244-7 (PDF)
doi.org/10.3390/books978-3-7258-2244-7

Contents

sensors

MDPI

Editorial

Sensors and Fault Diagnostics in Power System

Michał Kunicki [1,*], Jan Fulneček [2] and Pawel Rozga [3]

1 Department of Electrical Power Engineering and Renewable Energy, Opole University of Technology, 45-758 Opole, Poland
2 Department of Electrical Power Engineering, VSB Technical University of Ostrava, 708 00 Ostrava, Czech Republic; jan.fulnecek@vsb.cz
3 Institute of Electrical Power Engineering, Lodz University of Technology, Stefanowskiego 20, 90-537 Lodz, Poland; pawel.rozga@p.lodz.pl
* Correspondence: m.kunicki@po.opole.pl

The adequate technical condition assessment of key apparatuses is a crucial assumption in the delivery of reliable and continuous electric power to customers [1]. To meet this requirement, any fault in the power system must be detected and diagnosed as early as possible, with particular emphasis on the precision of the diagnostic process [2]. Various online and offline diagnostic methods are widely applied to the early detection of any system malfunctions [3]. Furthermore, a number of different sensors may also be applied to capture selected physical quantities that may be used to indicate the type of potential faults. A specific fault diagnostic process is typically performed by experts in the field; however, artificial intelligence (AI)-based systems are increasingly being proposed to support the decision-making process related to this task [4]. The essential step of the fault diagnostic process is signal analysis, supported by features including (but not limited to) signal processing, feature extraction, modeling, and prediction methods. Thus, the editors are pleased to present the collection of ten high-quality papers that deal with the main contemporary problems related to sensors and fault diagnostics in power system engineering. Hereunder, a brief review of each chapter of this book is presented.

In [5], a method of equivalent error is used as a criterion of the assessment of different algorithms applied for estimation of specific power system parameters. In order to indicate the best method for determining errors in measuring the synchronous parameters of the measured current or voltage waveforms, the authors propose a new form of a single error for all testing functions, which is called an equivalent error. This error is determined for each error value defined in the applicable standards for each of the selected 15 methods. According to the results, the use of the equivalent error algorithm can be very helpful in identifying a group of methods whose operation is satisfactory in terms of measurement accuracy for various types of disturbances (both in the steady state and in the dynamic state) that may occur in the power grid.

A problem of the accurate measurements and investigation of electromagnetic transients was raised in [6]. The authors proposed a capacitive electric field sensor-based measurement system to measure transient overvoltages in high-voltage substations. First, the concept and design of the measurement system is presented. Then, the design and concept are validated using tests performed in a high-voltage laboratory. Afterwards, two different calibration techniques are discussed: the simplified method (SM) and the coupling capacitance compensation (CCC) method. Finally, three recorded transients are evaluated using the calibration methods. The investigation revealed that the SM tends to overestimate the maximum overvoltage, highlighting the CCC method as a more suitable approach for calibrating transient overvoltage measurements. The proposed measurement system has been validated using various measurements and can be an efficient and flexible solution for the long-term monitoring of transient overvoltages in high-voltage substations.

Citation: Kunicki, M.; Fulneček, J.; Rozga, P. Sensors and Fault Diagnostics in Power System. *Sensors* **2024**, *24*, 5999. https://doi.org/10.3390/s24185999

Received: 11 September 2024
Accepted: 12 September 2024
Published: 16 September 2024

In order to address the challenges of low recognition accuracy and the difficulty in effective diagnosis in traditional converter transformer voiceprint fault diagnosis, a novel method is proposed in [7]. This approach considers the impact of load factors, utilizes a multi-strategy improved Mel-Frequency Spectrum Coefficient (MFCC) for voiceprint signal feature extraction, and combines it with a temporal convolutional network for fault diagnosis. In the first step, it improves the hunter–prey optimizer (HPO) as a parameter optimization algorithm and adopts IHPO combined with variational mode decomposition (VMD) to achieve denoising of voiceprint signals. Next, the preprocessed voiceprint signal is combined with Mel filters through the Stockwell transform. To adapt to the stationary characteristics of the voiceprint signal, the processed features undergo further mid-temporal processing, ultimately resulting in the implementation of a multi-strategy improved MFCC for voiceprint signal feature extraction. Simultaneously, load signal segmentation is introduced for the diagnostic intervals, forming a joint feature vector. Finally, by using the Mish activation function to improve the temporal convolutional network, the IHPO-ITCN is proposed to adaptively optimize the size of convolutional kernels and the number of hidden layers and construct a transformer fault diagnosis model. By constructing multiple sets of comparison tests through specific examples and comparing them with the traditional voiceprint diagnostic model, our results show that the model proposed in this paper has a fault recognition accuracy as high as 99%. The recognition accuracy was significantly improved, and the training speed also shows superior performance, which can be effectively used in the field of multiple fault diagnosis of converter transformers.

In [8], the authors presented an alternative approach to the Transformer Assessment Index (TAI) by proposing a relatively simple rating method called the Exploitation Perspective Index (EPI). The method provides two numerical indicators: the first reflects the overall technical condition of the particular unit, and the second shows the condition of the unit in the context of the entire fleet. The objective of the EPI method is to support the decision-making process regarding the technical condition assessment of each of the transformers in the target population, considering not only technical but also economic aspects of transformer maintenance. Application of the method is described step by step, including input data, parametrization of the weights, and interpretation of the output results it provides. The proposed method is evaluated by two representative use cases and compared with two other methods. As a result, EPI confirms its applicability, and it has already been successfully implemented by the electric power industry. EPI can be potentially freely adopted for any transformer fleet as well as for the specific situation of the utility by adjusting the relevant parameters.

One of the well-known problems related to smart grids are outliers. They can be generated in the power system due to aging system equipment, faulty sensors, incorrect line connections, etc. The existence of these outliers will pose a threat to the safe operation of the power system, reduce the quality of the data, affect the completeness and accuracy of the data, and thus affect the monitoring analysis and control of the power system. Therefore, timely identification and treatment of outliers are essential to ensure stable and reliable operation of the power system. In [9], the minorization–maximization algorithm was used to detect and localize the outliers and an estimation of unknown parameters of the Gaussian mixture model (GMM). According to the results, the proposed algorithm provides an effective method for the handling of outliers in the power system, which helps to improve the monitoring, analyzing, and controlling ability of the power system and to ensure the stable and reliable operation of the power system.

A simplified model of the Rogowski coil is proposed in [10]. In this paper, a new user-friendly model for Rogowski coils is presented and validated. The model's simplicity stems from utilizing information solely from the Rogowski coil datasheet. By establishing the input/output relationship, the output of the Rogowski coil is obtained. The effectiveness and accuracy of the proposed model are tested using both simulations and commercially available Rogowski coils. The results confirm that the model is simple, accurate, and easily implementable in various simulation environments for a wide range of applications and purposes.

System stability deterioration in microgrids commonly occurs due to unpredictable faults and equipment malfunctions. Recently, robust control techniques have been used in microgrid systems to address these difficulties. One of these alternative control strategies is proposed in [11]. The suggested approach can be used to maintain system stability in the presence of flaws, such as faulty actuators and sensors, as well as component failures. The proposed control is effective when the fault is never recognized (or when the fault is not being precisely known, and some ambiguity in the fault may be interpreted as uncertainty in the system's dynamics following the fault). The design is built around a derived sufficient condition in the context of linear matrix inequalities (LMIs) and the attractive ellipsoid technique. The ellipsoidal stabilization idea is to bring the state trajectories into a small region including the origin (an ellipsoid with minimum volume), and the trajectories will not leave the ellipsoid in the future. Finally, computational studies on a DC microgrid system are carried out to assess the effectiveness of the proposed fault-tolerant control approach. When compared with previous studies, the simulation results demonstrate that the proposed control technique can significantly enhance the reliability and efficiency of DC microgrid systems.

In [12], a new method for condition assessment of natural ester–mineral oil mixtures due to transformer retrofilling via sensing dielectric properties is announced. In this study, two accelerated aging processes were applied to mineral oil for 6 and 12 days to simulate mineral oil in service for 6 and 12 years. Moreover, these aged oils were mixed with 80% and 90% fresh natural ester oil. The dielectric strength, relative permittivity, and dissipation factor were sensed using an LCR meter and oil tester devices for all prepared samples to support the condition assessment performance of the oil mixtures. In addition, the electric field distribution was analyzed for a power transformer using the oil mixtures. Furthermore, the dynamic viscosity was measured for all insulating oil samples at different temperatures. From the obtained results, the sample obtained by mixing 90% natural ester oil with 10% mineral oil aged for 6 days is considered superior and achieves an improvement in dielectric strength and relative permittivity by approximately 43% and 48%, respectively, compared to fresh mineral oil. However, the dissipation factor was increased by approximately 20% but was at an acceptable limit. On the other hand, for the same oil sample, due to the higher molecular weight of the natural ester oil, the viscosities of all mixtures were at a higher level than the mineral oil.

A proposal of a novel test tool for diagnosis of contact resistance and measurement of selected types of conductive materials is presented in [13]. It is obvious that contact resistance is a fundamental criterion in the design of an electrical contact or contact system. The value of the contact resistance depends on the material used, the value of the applied force, the type of contact, and, last but not least, the quality of the surface and chemical layers. In this paper, an initial diagnosis of the contact material is performed based on the determination of the sample's specific resistivity by the four-wire method and the evaluation of the measurement uncertainty. The work is followed by the design of a testing device that uses crossed bars to measure the change in contact resistance as a function of the magnitude of the applied force. An analysis of the sample mounting method is performed here using FEM simulations of the current field and shows the interaction between the holder and the sample in terms of current line transfer. The proposed system is then used for experimental measurements of the material-dependent coefficient K_C for verification of existing or newly developed materials in electrical engineering, where the values of the K_C coefficient are not known.

The location of the grounding grid conductors is critical for performing corrosion diagnosis and maintenance work. An improved magnetic field differential method to locate the unknown grounding grid based on truncation errors and the round-off error analysis is presented in [14]. It was proven that a different order of the magnetic field derivative can be used to determine the position of the grounding conductor according to the peak value of the derivative. Due to the accumulative error of higher differentiation, the truncation error and rounding error were used to analyze the cumulative error and to determine the optimal step size to measure and calculate the higher differentiation. The possible range and probability distribution of the two kinds of errors at each order are described, and the index of peak position error was derived, which can be used to locate the grounding conductor in the power substation.

Conflicts of Interest: The authors declare no conflict of interest.

References

1. Kunicki, M.; Borucki, S.; Cichoń, A.; Frymus, J. Modeling of the Winding Hot-Spot Temperature in Power Transformers: Case Study of the Low-Loaded Fleet. *Energies* **2019**, *12*, 3561. [CrossRef]
2. Kunicki, M.; Cichon, A. Analysis on partial discharges variability in mineral oil under long-term AC voltage. *IEEE Trans. Dielectr. Electr. Insul.* **2018**, *25*, 1837–1845. [CrossRef]
3. Nagi, Ł.; Kozioł, M.; Kunicki, M.; Wotzka, D. Using a Scintillation Detector to Detect Partial Discharges. *Sensors* **2019**, *19*, 4936. [CrossRef] [PubMed]
4. Kunicki, M.; Wotzka, D. A Classification Method for Select Defects in Power Transformers Based on the Acoustic Signals. *Sensors* **2019**, *19*, 5212. [CrossRef] [PubMed]
5. Binek, M.; Rozga, P. Method of Equivalent Error as a Criterion of the Assessment of the Algorithms Used for Estimation of Synchrophasor Parameters Taken from the Power System. *Sensors* **2024**, *24*, 4619. [CrossRef] [PubMed]
6. Probst, F.L.; Beltle, M.; Tenbohlen, S. Measurement of Transient Overvoltages by Capacitive Electric Field Sensors. *Sensors* **2024**, *24*, 1357. [CrossRef] [PubMed]
7. Li, H.; Yao, Q.; Li, X. Voiceprint Fault Diagnosis of Converter Transformer under Load Influence Based on Multi-Strategy Improved Mel-Frequency Spectrum Coefficient and Temporal Convolutional Network. *Sensors* **2024**, *24*, 757. [CrossRef] [PubMed]
8. Kunicki, M.; Borucki, S.; Fulneček, J. Exploitation Perspective Index as a Support of the Management of the Transformer Fleet. *Sensors* **2023**, *23*, 8681. [CrossRef] [PubMed]
9. Qiao, L.; Gao, W.; Li, Y.; Guo, X.; Hu, P.; Hua, F. Smart Grid Outlier Detection Based on the Minorization–Maximization Algorithm. *Sensors* **2023**, *23*, 8053. [CrossRef] [PubMed]
10. Mingotti, A.; Betti, C.; Tinarelli, R.; Peretto, L. Simplifying Rogowski Coil Modeling: Simulation and Experimental Verification. *Sensors* **2023**, *23*, 8032. [CrossRef] [PubMed]
11. Soliman, H.M.; Bayoumi, E.H.E.; El-Sheikhi, F.A.; De Santis, M. Decentralized Sensor Fault-Tolerant Control of DC Microgrids Using the Attracting Ellipsoid Method. *Sensors* **2023**, *23*, 7160. [CrossRef] [PubMed]
12. Karaman, H.S.; Mansour, D.-E.A.; Lehtonen, M.; Darwish, M.M.F. Condition Assessment of Natural Ester–Mineral Oil Mixture Due to Transformer Retrofilling via Sensing Dielectric Properties. *Sensors* **2023**, *23*, 6440. [CrossRef] [PubMed]
13. Kacor, P.; Bernat, P.; Mlcak, T.; Hrabovsky, L. Realization of a Test Tool for Diagnosis of Contact Resistance and Measurement of Selected Types of Conductive Materials. *Sensors* **2023**, *23*, 5867. [CrossRef] [PubMed]
14. Yang, F.; Liu, S.; Lai, Y.; Hu, J.; Wang, S. An Improved Magnetic Field Method to Locate the Grounding Conductor. *Sensors* **2023**, *23*, 3879. [CrossRef] [PubMed]

sensors

MDPI

Article

An Improved Magnetic Field Method to Locate the Grounding Conductor

Fan Yang [1], Songlin Liu [1,*], Yijun Lai [1], Jiayuan Hu [2] and Shaohua Wang [2]

[1] State Key Laboratory of Power Transmission Equipment and System Security and New Technology, Chongqing University, Chongqing 400044, China
[2] State Grid Zhejiang Electric Power Research Institute, Hangzhou 310014, China
* Correspondence: liusonglin@cqu.edu.cn

Abstract: The location of the grounding grid conductors is critical for performing corrosion diagnosis and maintenance work. An improved magnetic field differential method to locate the unknown grounding grid based on truncation errors and the round-off errors analysis is presented in this paper. It was proven that a different order of the magnetic field derivative can be used to determine the position of the grounding conductor according to the peak value of the derivative. Due to the accumulative error of higher differentiation, the truncation error and rounding error were used to analyze to accumulative error and to determine the optimal step size to measure and calculate the higher differentiation. The possible range and probability distribution of the two kinds of errors at each order are described, and the index of peak position error was derived, which can be used to locate the grounding conductor in the power substation.

Keywords: magnetic field differential method; optimal step size; rounding error; truncation error

Citation: Yang, F.; Liu, S.; Lai, Y.; Hu, J.; Wang, S. An Improved Magnetic Field Method to Locate the Grounding Conductor. *Sensors* **2023**, *23*, 3879. https://doi.org/10.3390/s23083879

Academic Editors: Pawel Rozga, Michał Kunicki and Jan Fulneček

Received: 28 February 2023
Revised: 4 April 2023
Accepted: 7 April 2023
Published: 11 April 2023

1. Introduction

The grounding grid is an important piece of equipment that provides a common reference ground for various electrical equipment in the substation, quickly discharges the fault current in the event of a ground fault in the system, improves the ground potential distribution in the substation field, and ensures the safety of primary and secondary equipment and personnel under fault conditions [1,2]. The material of the grounding grid of the power system is mainly copper, which does not easily suffer from soil corrosion. In recent years, in Europe and the United States, steel gradually began to replace copper as the grounding grid material, but the steel grounding grid in operation for a relatively short period of time has not suffered serious corrosion. However, in China, India and other countries, since the conductors of the grounding grid are mostly made of steel, with the increase of the operation period, corrosion is prone to occur due to improper welding construction and the influence of geological conditions, and the conductors may become thinner or even broken [3–5], resulting in the grounding performance becoming reduced and the safety difficult to guarantee [6,7].

The grounding grid is buried in soil about 0.8 m deep underground, making it difficult to excavate and replace [8]. Therefore, predicting corrosion defects of the grounding grid based on information that can be obtained above the ground is important for guiding operation and maintenance [9–11] and has important engineering significance. In grounding grid research, generally, the optimization design of the grounding grid and the grounding performance of the grounding grid are studied [12,13]. In addition, many scholars have carried out research on grounding grid fault diagnosis, mainly based on electrical network theory [14–16] and electromagnetic field theory [17–29]. The method based on electrical network theory equivalently treats the branches of the grounding grid as pure resistances, injects current into the grounding grid, measures the conductor branch or node voltage of the grounding grid, studies and establishes a fault prediction diagnosis model, and uses optimization algorithms for solving [30–33]. These methods are based on the known topological structure of the grounding grid to perform grounding grid fault diagnosis. However, when diagnosing the grounding grid of a substation with a long history, sometimes the grounding grid drawing is missing or the actual structure of the grounding grid has a large error from the existing drawing due to reconstruction. In order to obtain the structure of the grounding grid, the grounding grid topology detection methods based on electromagnetic induction principle are widely used, mainly including the electrical source detection method (magnetic field method) [18–26] and the magnetic source detection method [27–29].

The magnetic source detection method uses a transmitting coil placed on the ground to pass a certain frequency of current. The induced currents generated by underground metal bodies under the excitation of the primary magnetic field produce changing magnetic fields in the surrounding space, called secondary magnetic fields. The position and orientation of underground metal conductors can be obtained by measuring the secondary magnetic field information received by the surface receiving coils. Magnetic source detection methods are divided into the frequency domain electromagnetic method and the time domain electromagnetic method according to different excitation and response characteristics. The time domain electromagnetic method is also called the transient electromagnetic method [27]. However, detection methods based on magnetic sources are greatly affected by metal structures.

The magnetic field method injects a sinusoidal current of a specific frequency into the grounding grid through the two upper guide wires of the grounding grid, measures the magnetic induction intensity generated by the current-carrying conductor of the grounding grid on the earth's surface, and analyzes the distribution characteristics and laws of the magnetic induction intensity to determine the structure and fault states of the grounding grid [25]. Knowing the distribution of the magnetic field generated by the conductor of the current-carrying grounding grid on the surface, the inverse problem equation of the magnetic field can be established to solve the topological structure of the grounding grid. However, the inverse problem is usually ill-conditioned, it is difficult to obtain a unique solution or a stable solution, and complex regularization is required. The analysis method based on the magnetic field differential can avoid the solution of the inverse problem [34–37], but the error introduced in the numerical differential calculation may make the position of the grounding grid conductor deviate, and the selection of an appropriate measurement step is of great importance for the accurate determination of the conductor position.

In order to improve the reliability of the magnetic field differential method, this paper analyzed the range and probability distribution of the truncation error and rounding error produced by the numerical differential of the magnetic field method, derived the expressions of the mathematical expectation and variance of the differential main peak position error, and finally used Monte Carlo simulation to calculate the local optimal measurement step size under the second and fourth order differentials so that the magnetic field differential method reduced the total error and improved the accuracy of conductor positioning.

2. Error Analysis of Magnetic Field Differential Method

2.1. Magnetic Field Differentiation Method

The grounding grid is composed of regularly connected horizontal grounding conductors. The branch position information can be obtained by analyzing the distribution of the magnetic field generated by the current-carrying conductor branch. The rough outline of the grounding grid topology can be obtained by measuring the magnetic flux density distribution on the ground surface. However, due to the wide influence range of the magnetic field and the influence of the superposition of the magnetic field, it is not possible to accurately locate the conductor using the original magnetic field distribution characteristics. The magnetic field differential method can enhance the peak characteristics of the magnetic field through high-order differential and can improve the positioning accuracy of the grounding grid conductor. The magnetic field differentiation method is described below using a single current-carrying conductor model.

The infinitely long conductor is placed on the x-axis through the coordinate origin, as shown in Figure 1; the conductor is buried horizontally in a single layer of uniform soil with a magnetic permeability μ; the buried depth is h; the current flowing through the conductor is I; and the direction of the current is vertical outward in the y-z plane. For point P on the ground surface, the vertical distance from the current-carrying conductor is ρ, and the angle between the line segment OP and the z-axis is θ.

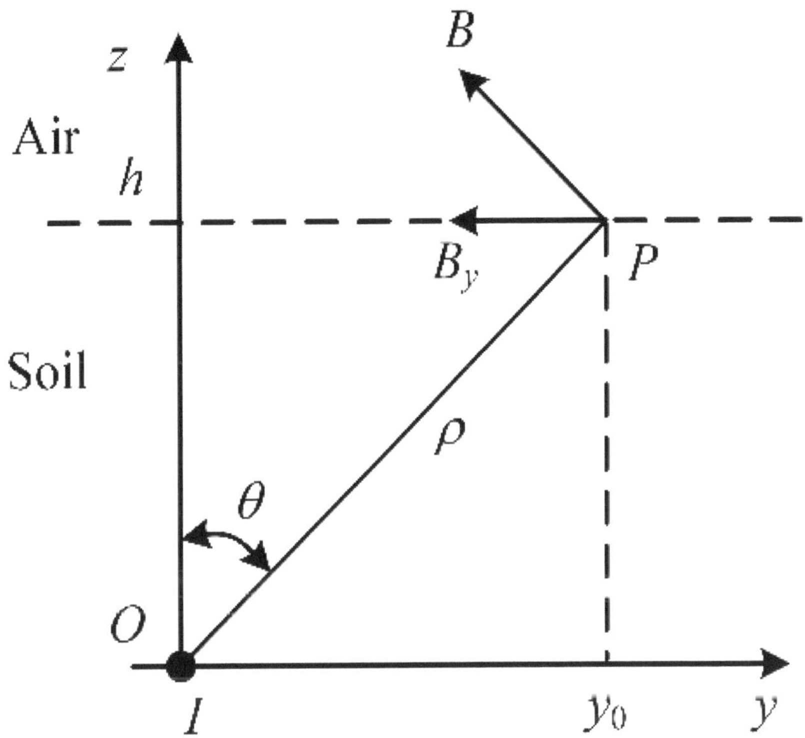

Figure 1. Single conductor current-carrying model.

According to the principle of potential continuity, the potential on both sides of the interface between the conductor and the soil is equal, and the resistivity of the conductor is significantly smaller than that of the soil. Therefore, the current density in the soil is significantly smaller than that inside the conductor, and the influence of the soil leakage current is negligible. Neglecting the leakage current of the conductor in the soil, the magnetic flux density generated by the current-carrying conductor at point P can be expressed by Ampere's loop theorem as follows:

$$\boldsymbol{B} = \frac{\mu I}{2\pi\rho}\mathbf{e}_{\phi}. \tag{1}$$

where $\mathbf{e}_{\phi}$ is the unit vector in the direction perpendicular to the unit vector $\mathbf{e}_{\rho}$, and $\mathbf{e}_{\rho}$ is the unit vector in the direction of the line OP.

From the geometric relationship, it can be concluded that the magnetic flux density $B_y(y)$ parallel to the ground generated by the current-carrying conductor at point P is:

$$B_y(y) = -\frac{\mu I h}{2\pi}\frac{1}{h^2 + {y_0}^2}. \quad (2)$$

Equation (2) describes the distribution of magnetic flux density in the horizontal direction generated by a single current-carrying conductor, which is called the shape function. For a grid-shaped grounding grid, the magnetic flux density distribution in the horizontal direction on the ground surface of the grounding grid can be equivalent to the superposition of the shape functions of each current-carrying branch of the grounding grid.

The higher the differential order of the shape function, the more complex its expression. Considering that the even-order derivative has the main peak characteristic, this paper calculated only the second-order differential and fourth-order differential of the shape function.

$$B_y^{(2)}(y) = \frac{\mu I h}{\pi}\frac{h^2 - 3y^2}{(h^2 + y^2)^3}, \quad (3)$$

$$B_y^{(4)}(y) = -\frac{12\mu I h}{\pi}\frac{h^4 - 10h^2y^2 + 5y^4}{(h^2 + y^2)^5}. \quad (4)$$

When I = 1 A, h = 1 m, the curves of the shape function $B_y(y)$, the second-order differential $B_y^{(2)}(y)$, and the fourth-order differential $B_y^{(4)}(y)$ of the shape function are shown in Figure 2.

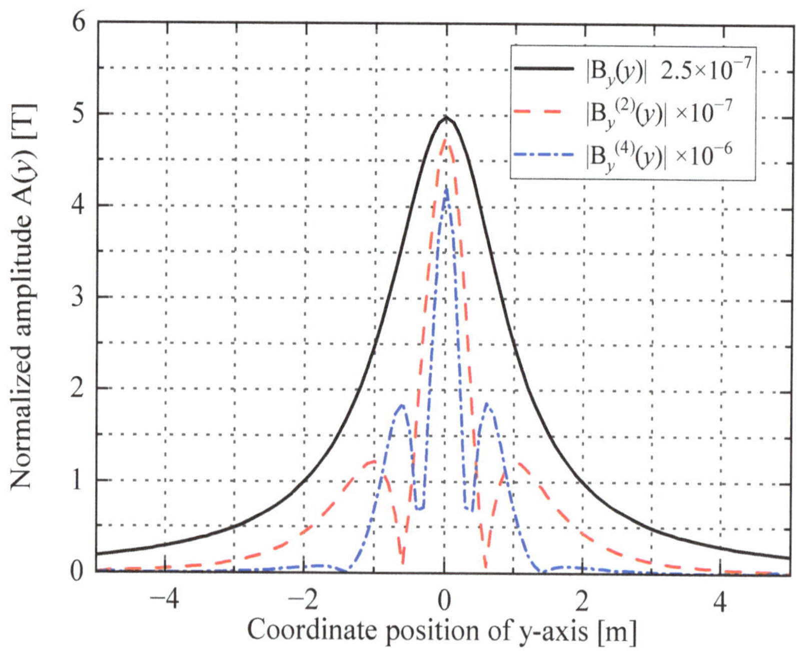

Figure 2. Curves of three shape functions.

The main peak width in Table 1 is the width between the two zero points (or 1% of the main peak value) of the main peak. The side peak width is the width between the two zero points (or 1% of the main peak value) of the side peak adjacent to the main peak.

The Widess resolution R_W is the ratio of the energy of the main peak maximum b_M^2 of the function to the total energy of the function E:

$$R_W = \frac{b_M^2}{E}, \quad (5)$$

where b_M is the maximum of the shape function

$$E = \int_{-\infty}^{\infty} b^2(y)dy. \quad (6)$$

where b is the shape function.

Table 1. Comparison of function shape properties.

Function	Main Peak Width (m)	Side Peak Width (m)	Widess Resolution
$\lvert B_y(y) \rvert$	19.90	-	0.6361
$\lvert B_y^{(2)}(y) \rvert$	1.1552	3.3739	1.6849
$\lvert B_y^{(4)}(y) \rvert$	0.6504	1.0516	2.2847

From the comparison of the data in Table 1, it can be seen that, with the second-order and fourth-order derivatives of $B_y(y)$, the width of the main peak and the width of the side peaks gradually decrease, the total number of peaks and the Widess resolution gradually increase, and the signal recognition ability enhances. According to Equations (4) and (5), the positions of the main peaks of functions $B_y^{(2)}(y)$ and $B_y^{(4)}(y)$ are the same as those of the current-carrying conductors and are both at $y = 0$. Therefore, the positions of the main peaks of the second-order derivatives or fourth-order derivatives of the magnetic flux density $B_y(y)$ can be used to determine the locations of the grounding grid branches in the measurement area and thus to map the grounding grid topology.

2.2. Simulation of Current-Carrying Grounding Gird

As shown in Figure 3, a 2 × 2 grid of flat steel (cross-sectional area of 4 cm × 3 mm) was laid with a grid spacing of 5 m. The current of 1 A was injected from node 4 and flowed out from node 3. A Cartesian coordinate system x-y-z was established with node 1 as the origin of the coordinate axis, and the positive direction of the z-axis was perpendicular to the x-y plane upward. Below the plane $z = h$, there was a single layer of homogeneous soil with magnetic permeability μ. The magnetic permeability of the soil was approximated by taking the permeability μ_0 in a vacuum. The resistivity of the conductor was significantly larger than that of the soil, and the soil leakage current had a negligible effect on the simulation results of the magnetic flux density. Therefore, we set a typical value of 80 Ω·m for the soil resistivity.

Through the shape function of a single current carrying conductor, it can be known that the peak values of the horizontal components B_x and B_y of the magnetic flux density can reflect the conductor position in different directions; thus, the modulus of the magnetic flux density is more important than its direction for the positioning of the conductors. $\lvert B_x \rvert + \lvert B_y \rvert$ can reflect all current-carrying conductors in the x-y direction. In order to study the magnetic flux density generated by the current-carrying grounding grid, the simulation was performed using MATLAB based on the finite element method. The magnetic flux density $\lvert B_x(x,y) \rvert + \lvert B_y(x,y) \rvert$ generated by the current-carrying grid branch was detected in the horizontal plane at a distance of $h = 0.5$ m from the x-y plane, while $\lvert B_y(y) \rvert$ was detected on the survey line at the position $x = 6$ m, and the results are shown in Figures 4 and 5.

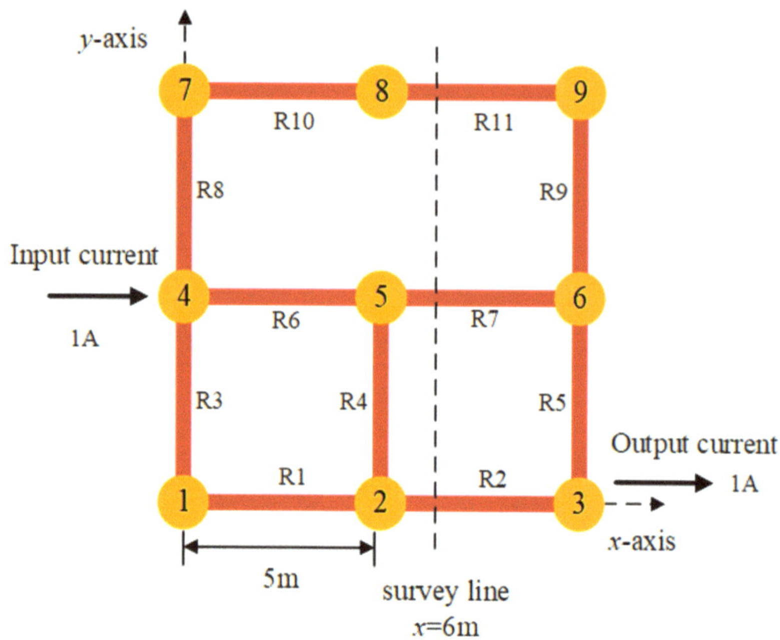

Figure 3. Simple current-carrying grid model.

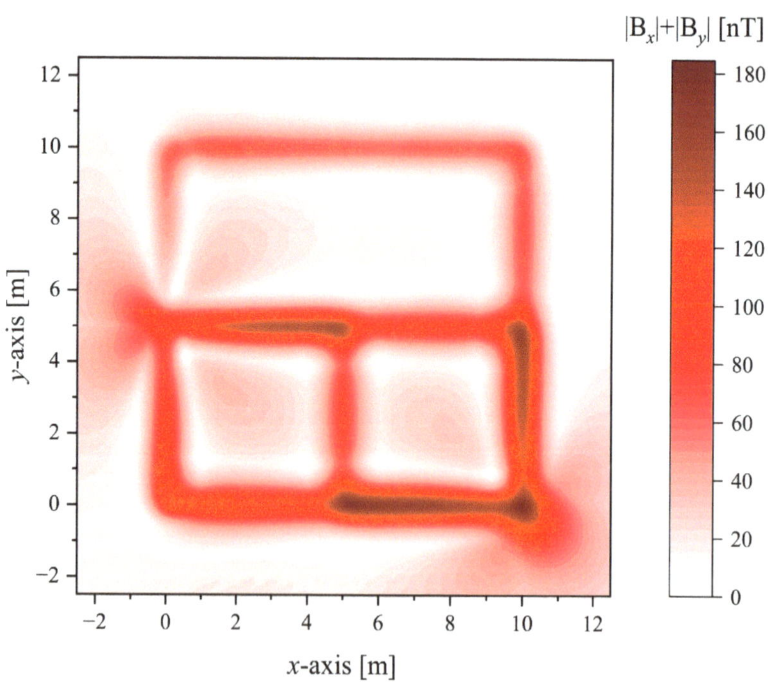

Figure 4. The distribution of the magnetic flux density mode at the plane h = 0.5 m.

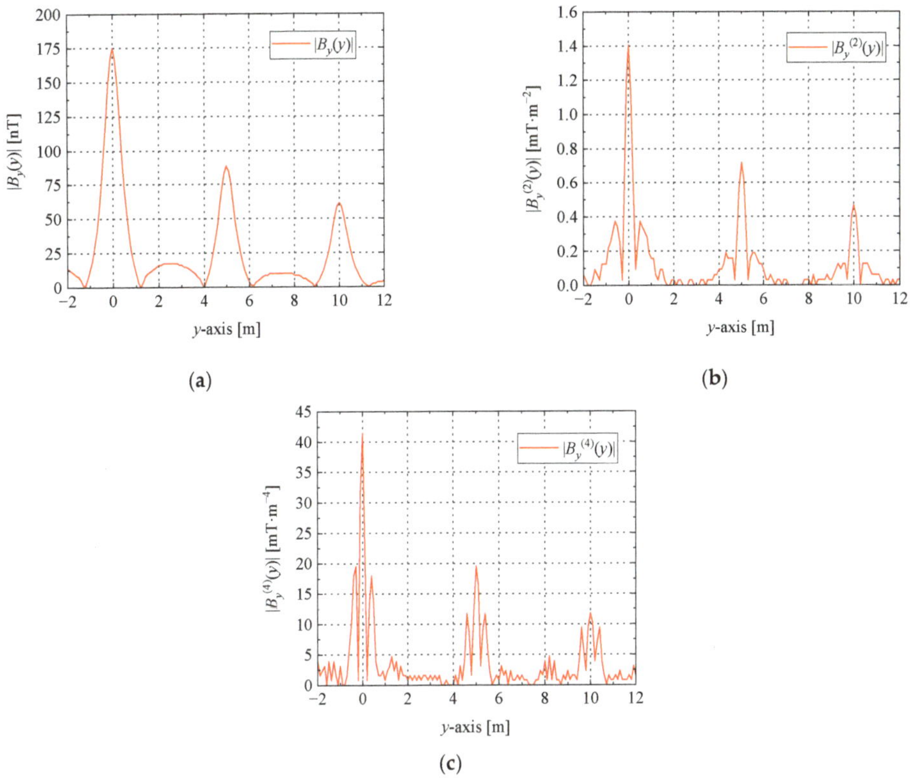

Figure 5. The curve of magnetic flux density differential at the survey line $x = 6$ m: (**a**) the absolute value curve of magnetic flux density; (**b**) the second-order differential absolute value curve of the magnetic flux density; (**c**) the fourth order differential absolute value curve of the magnetic flux density.

The results of the simulation are shown in Table 2. The grid spacing defined in the simulation is 5 m, and the errors of the grid spacing according to functions $|B_y(y)|$, $|B_y^{(2)}(y)|$, and $|B_y^{(4)}(y)|$ are 2.18%, 0.75%, and 0.28%, respectively.

Table 2. Simulation results of grid size.

Function	Calculated Grid Spacing (m)	Side Peak Width (m)
$|B_y(y)|$	4.8912	2.18
$|B_y^{(2)}(y)|$	4.9623	0.75
$|B_y^{(4)}(y)|$	4.9861	0.28

2.3. Numerical Differential Error Analysis

When performing the position measurement of the grounding grid conductor, due to the limitations of the measurement equipment size, measurement time consumption, and other factors, it was not possible to measure a sufficient number of data points at the substation; thus, the horizontal component of the magnetic flux density measured on a certain survey line was a discrete sequence. When the differential method is used to locate the grounding conductor, a numerical calculation method is required. However, since the measured information did not contain the function expression and the noise introduced in the measurement process was unavoidable, the calculation result obtained by the difference quotient did not have high reliability, and the accuracy of the difference quotient result depends on the step size of the difference, i.e., the measurement interval. Sometimes, a small difference step size may lead to a large calculation error [38]. Therefore, it is necessary to analyze the sources of error in the process of numerical differentiation. In the following, the error situation of the magnetic field differential is analyzed from the perspective of truncation error and rounding error.

The formula for the central difference quotient commonly used In numerical differentiation can be expressed as:

$$G(x) = \frac{f(x+d) - f(x-d)}{2d}. \tag{7}$$

where d is the differential step size.

Substituting $f(x_0 \pm d)$ into formula (7) after performing Taylor expansion at $x = x_0$, the truncation error is:

$$\left|f'(x_0) - G(x_0)\right| \leq d^2 M/6, \tag{8}$$

where $M \geq \max\limits_{|x-a| \leq d} \left|f^{(3)}(x)\right|$, and a is the center of the interval over which the maximum value of $f^{(3)}(x)$ is taken.

From the perspective of truncation error, the smaller the step size d, the more accurate the calculation result of numerical differentiation.

Considering the truncation error in the process of magnetic field differentiation, for the current-carrying grid shown in Figure 3, the survey line $x = x_0$ is selected, the coordinates of the measurement starting point are (x_0, y_0), and the position of the k-th measurement point along the survey line on the y-axis is noted as $y_k = y_0 + kd$, where d is the measurement interval. The nth order difference quotient of the y-direction component of the magnetic flux density $B_y(y)$ can be expressed as:

$$B_t{}^{(n)}(y_k) = \frac{B_t{}^{(n-1)}(y_{k+1}) - B_t{}^{(n-1)}(y_{k-1})}{2d}, \tag{9}$$

where the order $n \geq 1$, when $n = 1$, $B_t(y_k) = B(y_k)$.

The truncation error due to numerical differentiation can be expressed as the difference between the difference quotient and the differential quotient:

$$E_t^{(n)}(y_k) = B_t{}^{(n)}(y_k) - B^{(n)}(y_k). \tag{10}$$

When performing numerical difference calculations, the difference of two approximately equal numbers can result in a significant loss of valid numbers. The input point $f(x+d)$ of the difference quotient is denoted as $\hat{f}(x+d)$, and the error between the input point $\hat{f}(x+d)$ and the real value $f(x+d)$ is denoted as ε.

$$G(x) = \frac{\hat{f}(x+d) - \hat{f}(x-d)}{2d}. \tag{11}$$

The error between the real value of the nth-order derivative $f^{(n)}(x)$ of function $f(x)$ and its numerical calculation result can be expressed as:

$$\delta(f'(x)) = f'(x) - G(x) = \frac{\varepsilon_2 - \varepsilon_1}{2d} + \frac{h^2}{6} f^{(3)}(a). \tag{12}$$

where $a \in (x-d, x+d)$. If $\varepsilon_m = \max\{|\varepsilon_1|, |\varepsilon_2|\}$ is defined, the upper bound $E_{r\max}$ of the rounding error can be expressed as:

$$E_{r\max} = \frac{\varepsilon_m}{d}. \tag{13}$$

Therefore, from the perspective of rounding error, a step size d that is too small will result in a large rounding error.

2.4. Peak Position Error Analysis

In order to accurately locate the grounding grid conductor, it is necessary to determine an optimal measurement interval d. The mathematical expectation $M(d)$ and variance $D(d)$ of the peak deviation can be used to describe the degree of the conductor positioning error, which is derived below.

First, consider the impact of the randomness of the measurement points on the position error of the main peak. The survey line $x = x_0$ is selected, the coordinates of the measurement starting point are (x_0, y_0), and the position of the k-th measurement point along the survey line on the y-axis is noted as $y_k = y_0 + kd$, where d is the measurement interval. The probability density function of the truncation error $E_t(y_k)$ at y_k is $p_t = 1/d$; thus, the randomness of $E(y_k)$ is determined by the rounding error.

Let the probability density of the rounding error E_r at a measurement point be p_r, and p_r under the same measurement step d and order n is not affected by the location of the measurement point; thus, p_r is a function of E_r, where E_r is a uniformly distributed random variable on $[-E_{r\max}, E_{r\max}]$. Then:

$$p_r(E_r) = \begin{cases} \frac{1}{2E_{r\max}}, & |E_r| < E_{r\max} \\ 0, & |E_r| > E_{r\max} \end{cases}. \tag{14}$$

At the n-th order, E_r is obtained by accumulating the two rounding errors E_{r1} and E_{r2} of the n-1 order, and then $p_r^{(n)}(E_r = \frac{E_{r_1}+E_{r_2}}{2d})$ can be expressed as:

$$p_r^{(n)}(E_r) = \int_{-\infty}^{\infty} 2d \cdot p^{(n-1)}(E_{r_1}) p^{(n-1)}(2dE_r - E_{r_1}) \mathrm{d}E_{r_1}. \tag{15}$$

When n = 1~4, the distribution of $p(n)$ r with respect to E_r is shown in Figure 6.

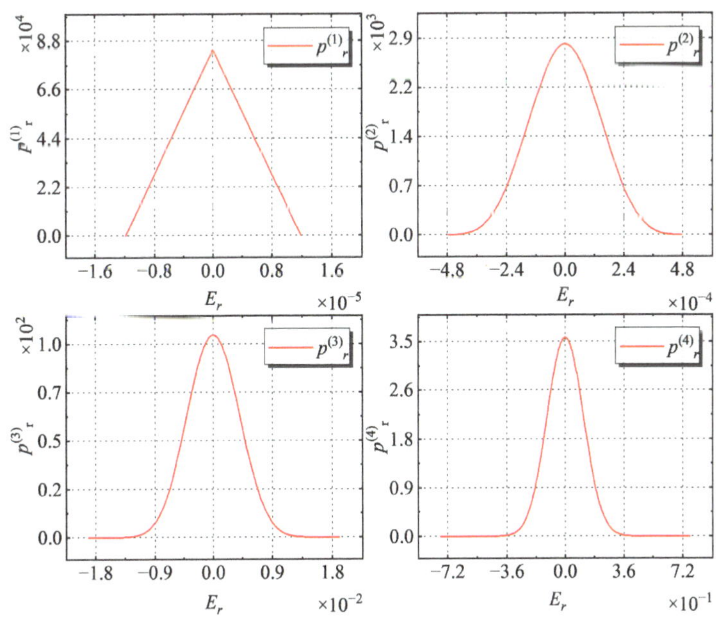

Figure 6. When n = 1~4, the distribution of $p_r^{(n)}$ on E_r.

The total error $E(y_k)$ is the sum of the truncation error and rounding error; thus, the probability density function $p(E(y_k))$ at y_k can be expressed as:

$$p(E(y_k)) = p_t p_r(E_r(y_k)) = \frac{p_r(E(y_k) - E_t(y_k))}{d}. \tag{16}$$

The error at y_k is $E(y_k)$ and is denoted as event k; each event k is independent of each other, and the probability density function of events 1, 2, ... , m being established at the same time is $\prod_{k=1}^{m} p(E(y_k))$, where the value range of $E(y_k)$ is as follows:

$$E_t(y_k) - E_{r\max}(y_k) \leq E(y_k) \leq E_t(y_k) + E_{r\max}(y_k). \tag{17}$$

The peak position of the measured and calculated value $B_C(y)$ of the n-th order derivative of the magnetic flux density is denoted as y_{pC}. Due to the presence of errors, there is a deviation between y_{pC} and the peak position y_{pR} of the real value $B_R(y)$. The deviation Δy_p is the difference between the calculated value and the real value, i.e., $\Delta y_p = y_{pC} - y_{pR}$. The probability density function $\prod_{k=1}^{m} p(E(y_k))$ is used to derive the mathematical expectation of the main peak position error.

First, consider the mathematical expectation of the offset caused by the error E_1 at y_1 when the errors of other points are constant. Assuming that E_1 takes only s discrete values, E_{10}, $E_{10} + \Delta E_1$, $E_{10} + 2\Delta E_1$, ... , $E_{10} + (s-1)\Delta E_1$, it corresponds to the distribution of a total of s function errors on y, and the cumulative expectation is:

$$\sum_{j=1}^{s} p(E_{10} + (j-1)\Delta E_1)\Delta y_{p1j}\Delta E_1. \tag{18}$$

Extending E_1 from the discrete distribution to the case of continuous distribution, i.e., $s \to \infty$, the above formula becomes:

$$\lim_{s\to\infty}\sum_{j=1}^{s} p(E_{10} + (j-1)\Delta E_1)\Delta y_{p1j}\Delta E_1 = \int_{E_{\max-}(y_1)}^{E_{\max+}(y_1)} p(E_1(y_1))\Delta y_{p1} dE_1 \tag{19}$$

Equation (19) represents the mathematical expectation of the offset due to the error E_1 at y_1 when the error at other points is constant.

$Y_q = p(E(y_1))\Delta y_{p1}$. $\xi(y_1)$ is the result of integrating $\prod_{q=1}^{m} p(E(y_q))\Delta y_p$ over $E(y_q)$, i.e.,

$$\xi(y_1) = \int_{E_{\max-}(y_m)}^{E_{\max+}(y_m)} \cdots \int_{E_{\max-}(y_1)}^{E_{\max+}(y_1)} Y_q \mathrm{d}E(y_1)\cdots \mathrm{d}E(y_m) \tag{20}$$

Equation (20) represents the expectation obtained by taking all possible cases of $E(y_1)$, $E(y_2)$, ... , $E(y_m)$ at y_1.

Notice that Δy_p is affected by all y_i; thus, the above equation cannot be expressed in terms of $\prod_{i=1}^{m} \int_{E_{\max-}(y_1)}^{E_{\max+}(y_1)} Y_q \mathrm{d}E(y_1)$.

Finally, consider that y_1 can be varied within the first interval segment $[y_{\min}, y_{\min} + d)$. In summary, the mathematical expectation $M(d)$ of the main peak deviation can be expressed as:

$$M(d) = \int_{y_{\min}}^{y_{\min}+d} \xi(y_1)\mathrm{d}y_1 \tag{21}$$

At the same time, the degree of dispersion of Y_q should also be considered; thus, the variance *D*(*d*) of the main peak position error caused by d can be expressed as:

$$D(d) = \int_{y_{\min}}^{y_{\min}+d} \int_{E_{\max-}(y_m)}^{E_{\max+}(y_m)} \cdots \int_{E_{\max-}(y_1)}^{E_{\max+}(y_1)} (Y_q - M(d))^2 dy_1 \cdots dy_m dx_1 \quad (22)$$

3. Experimental Analysis

3.1. Simulation Experiment

The time complexity of the algorithm used to calculate the main peak position error expectation *M*(*d*) and the main peak position error variance *D*(*d*) is exponential time complexity O(k^n). It is difficult to obtain results quickly through computation, but the constructed and described stochastic process and probability distribution are completely accurate models; thus, the Monte Carlo simulation can be used to obtain approximate results of the problem. The method of the Monte Carlo simulation increases the number of trials, and if the test results converge when the number of trials is sufficient, the final converged value is used as the simulation result.

First, the mean peak position error *M*(*d*) and the variance of the peak position error *D*(*d*) were calculated at a given measurement step *d*. Taking the current-carrying grid in Figure 3 as an example, the selected survey line was x = 6 m and the measurement step *d* was 0.05 m. The expectation *M*(*d*) and the variance *D*(*d*) of the main peak position error at the second-order differential and the fourth-order differential for the three conductors (y = 0 m, y = 5 m, y = 10 m) were calculated using Monte Carlo simulations, and the results are shown in Figure 7.

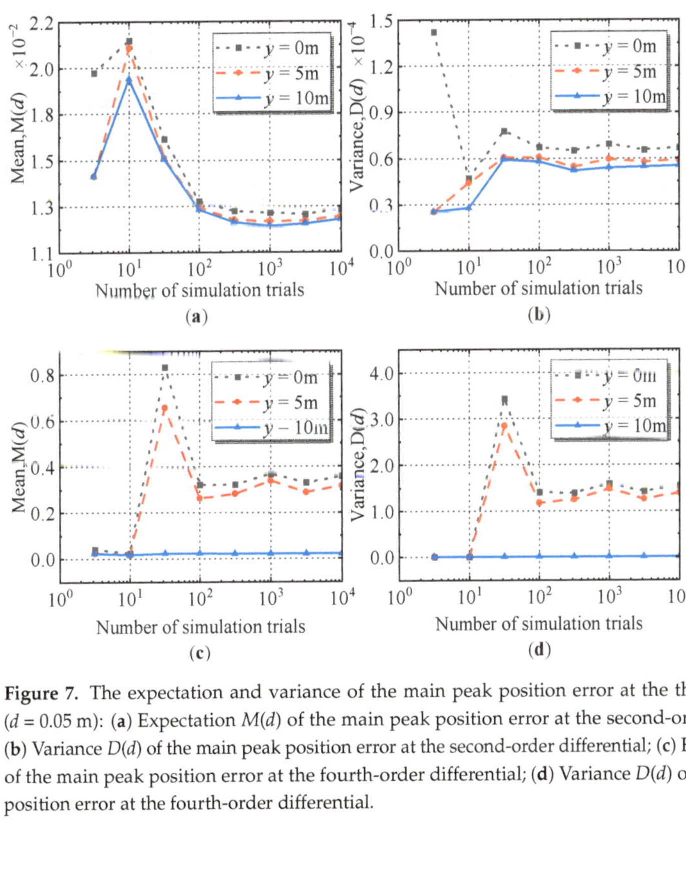

Figure 7. The expectation and variance of the main peak position error at the three conductors (*d* = 0.05 m): (**a**) Expectation *M*(*d*) of the main peak position error at the second-order differential; (**b**) Variance *D*(*d*) of the main peak position error at the second-order differential; (**c**) Expectation *M*(*d*) of the main peak position error at the fourth-order differential; (**d**) Variance *D*(*d*) of the main peak position error at the fourth-order differential.

It can be seen from Figure 7 that the error expectation and variance of the main peak position at the three conductor positions (y = 0 m, y = 5 m, y = 10 m) gradually tended to converge as the number of simulation tests increased; thus, the convergence value can be used as the expectation and variance of the main peak position error of the simulation.

Then, Monte Carlo simulations were used to calculate the expectation $M(d)$ and variance $D(d)$ of the main peak position error at three conductor locations (y = 0 m, y = 5 m, y = 10 m) with different measurement steps along the survey line x = 6 m, where the measurement step d takes values from 0.01 m to 1 m. When the number of simulation experiments was 10^4 times, the expectation and variance of the main peak position error for the second-and fourth-order differentials of the magnetic field at each measurement step d are shown in Figure 8.

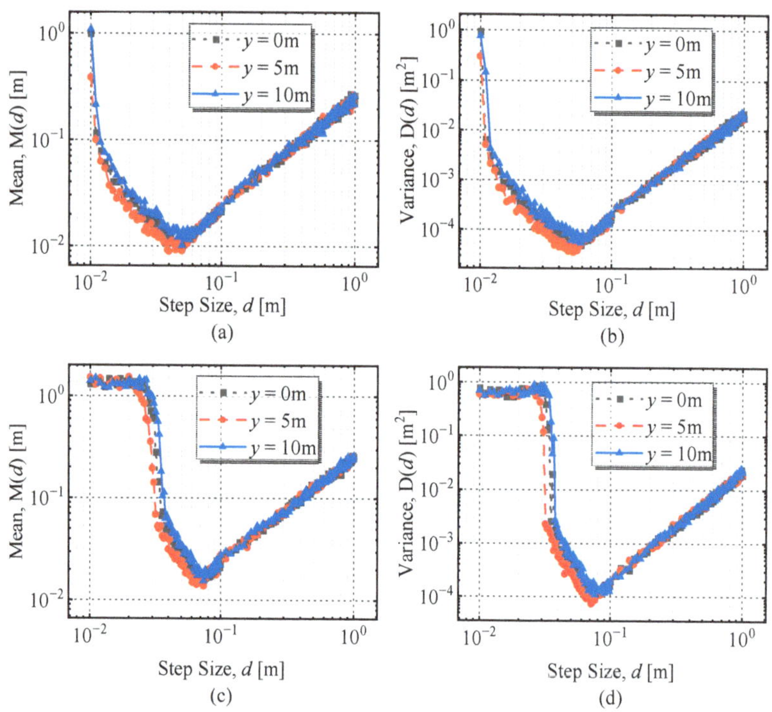

Figure 8. The expectation and variance of the main peak position error at the three conductors (d = 0.01~1 m): (**a**) Expectation $M(d)$ of the main peak position error at the second-order differential; (**b**) Variance $D(d)$ of the main peak position error at the second-order differential; (**c**) Expectation $M(d)$ of the main peak position error at the fourth-order differential; (**d**) Variance $D(d)$ of the main peak position error at the fourth-order differential.

According to Figure 8, the second-order differentiation, the main peak position error expectation, and variance diverged when d was close to 0, and they reached the minimum when d = 0.04 m and gradually increased with the increase of d. At this time, the local optimum value of the measurement step was 0.04 m. Similarly, when the order was 4, the local optimal value of the measurement step was 0.06 m.

Finally, to confirm the generalizability of the locally optimal step size derived from the x = 6 m survey line, several survey lines were selected at different locations to calculate the main peak position error expectation $M(d)$ and variance $D(d)$ using the Monte Carlo simulation. The survey line x_0 was taken in the range of [0 m, 10 m], and each time x_0 was increased by 2 m. The calculation results are shown in Table 3.

Table 3. Local optimal step size at different survey line positions.

x_0/m	0	2	4	6	8	10
Step size (2nd)	0.04	0.04	0.03	0.03	0.03	0.03
Step size (4th)	0.06	0.05	0.05	0.05	0.05	0.05

If the magnetic field differentiation method is used to locate the grounding grid conductors at the power substation and it is desired to minimize the impact of the numerical differential calculation error on the conductor location, an optimal measurement step needs to be selected. According to the results in Table 3, the measurement step size d = 0.04 m or d = 0.05 m can be selected as the optimal measurement step size.

3.2. Substation Field Experiment

In order to verify the effectiveness of the optimal step size selection, a field experiment of substation grounding grid conductor positioning was carried out. In the experiment, three different measurement steps were used to measure the magnetic flux density on the surface along the survey line, and the differential method was used to calculate the second-order and fourth-order differentials of the magnetic flux density at each step.

The schematic diagram of the conductor positioning experiment is shown in Figure 9a. The current output of the excitation source was connected to grounding lead conductors at the diagonal position in the test area. Eight PCB coils formed an array of magnetic field sensors with adjustable spacing. The signal conditioning circuit filtered and amplified the output signal of the magnetic field sensor. An 8-channel, 24-bit ADC (ADS1278) was used to convert the analog signal from the signal conditioning circuit into a digital signal. The ESP32 microcontroller communicated with the ADC through the SPI interface and sent the converted digital signal to the laptop via Wi-Fi. A photo of the experimental site is shown in Figure 9b.

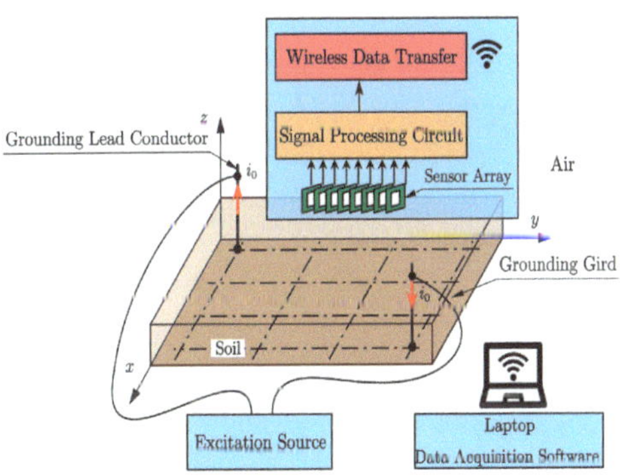

(a)

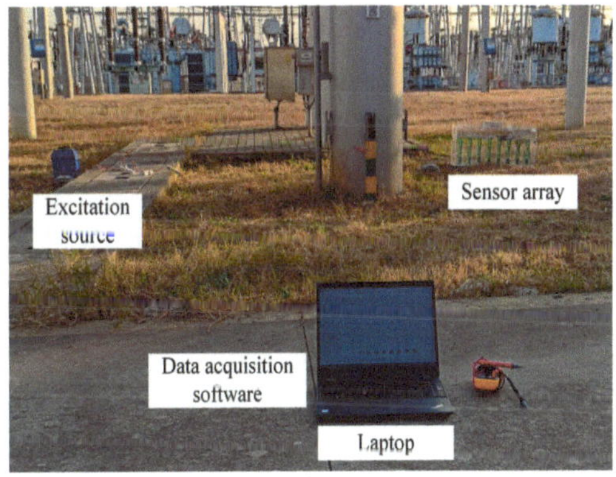

(b)

Figure 9. The 500 kV substation field experiments: (**a**) schematic diagram of conductor positioning experiment; (**b**) field experiment photo.

Figure 10 shows a schematic diagram of survey lines and the topology of part of the substation's grounding grid. The spacing of the ground grid conductors in the experimental area was unevenly distributed. The measurement area is shown in Figure 10, where the red dotted lines indicate the survey lines. Ten survey lines were laid out with a line spacing of 0.5 m. Each line was 20 m long, with a measurement point spacing of 0.02 m, 0.05 m, and 0.10 m. The sensing array was measured along the survey lines to record data at all locations. Two conductors, A and B, whose approximate positions were known, were located below the survey line.

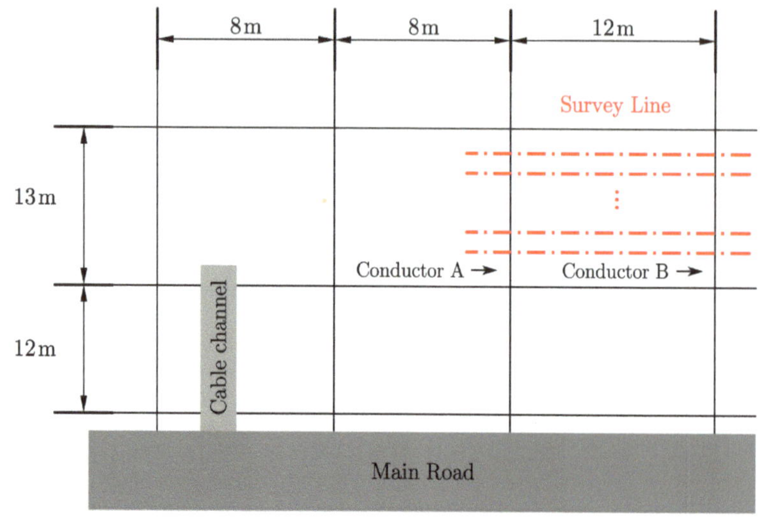

Figure 10. Schematic diagram of survey lines and topology.

Figure 11a shows the normalized absolute value of the magnetic flux density measured along the survey line when the measurement step length $d = 0.02$ m. Figure 11b,c shows the normalized absolute values of the second- and fourth-order differentials of the magnetic flux density calculated from the measurement results, respectively. It can be seen from the figure that the calculation result of the second-order differential can improve the Widess resolution and maintain the peak characteristic of the conductor position. However, due to the influence of numerical differential error and random noise in the measurement process, the calculated value of the fourth-order differential did not have good peak characteristics, and it is difficult to distinguish the position information of the conductor from the result.

Figure 12a shows the normalized absolute value of the magnetic flux density measured along the survey line when the measurement step length $d = 0.05$ m. Figure 12b,c shows the normalized absolute values of the second- and fourth-order differentials of the magnetic flux density calculated from the measurement results, respectively. It can be seen from the figure that the calculation results of the second-order differential and fourth-order differential can both improve the Widess resolution and maintain a good conductor position peak characteristic, and the calculation results reflect the actual position of the conductor.

Combined with the conclusion in Section 3.1, the local optimal measurement step d can be selected as 0.04 m or 0.05 m. Comparing Figures 11c and 12c, it was found that, due to the influence of rounding errors, the fourth derivative of the magnetic flux density when the measurement step $d = 0.02$ m, compared with $d = 0.05$ m, the peak characteristic was completely lost, and the conductor was difficult to locate through the differential result. Comparing Figures 11b and 12b, it was found that, due to the influence of rounding errors, the peak characteristic fluctuation of the second derivative of the magnetic flux density was more obvious when the measurement step length $d = 0.02$ m compared with $d = 0.05$ m, and the impact of measurement errors on the calculation results was more significant. This result is consistent with the conclusion stated in Section 3.1.

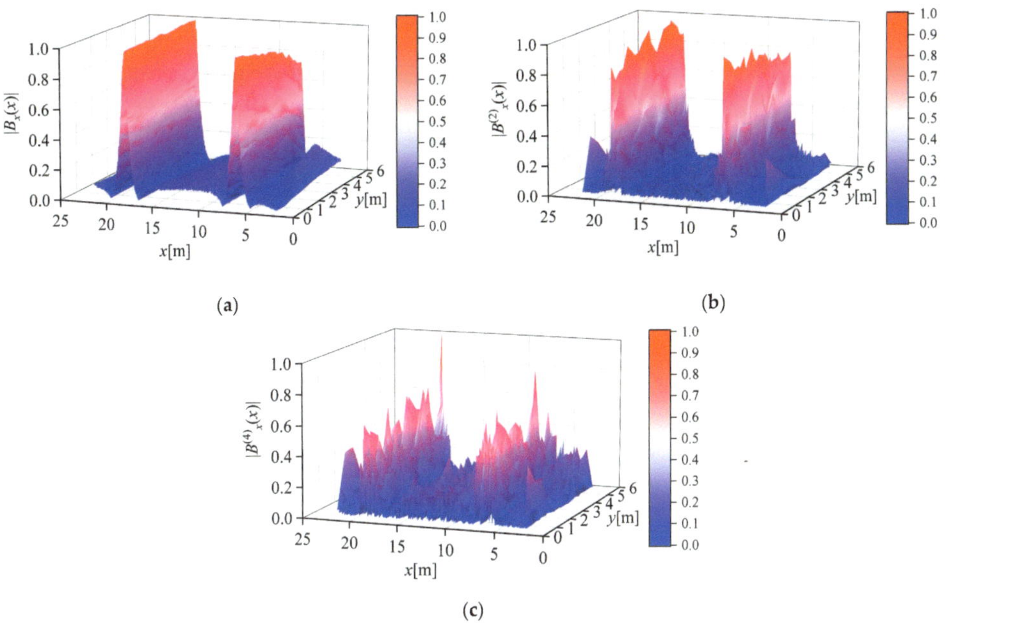

Figure 11. Normalized magnetic flux density absolute value measurement results when the measurement step size $d = 0.02$ m: (**a**) magnetic flux density; (**b**) second-order differential of the magnetic flux density; (**c**) fourth-order differential of the magnetic flux density.

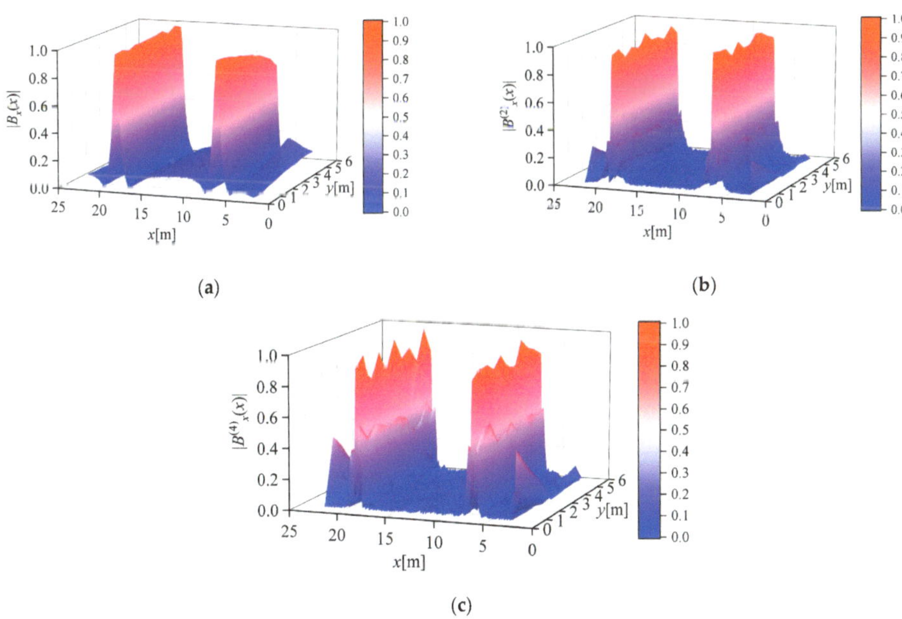

Figure 12. Normalized magnetic flux density absolute value measurement results when the measurement step size $d = 0.05$ m: (**a**) magnetic flux density; (**b**) second-order differential of the magnetic flux density; (**c**) fourth-order differential of the magnetic flux density.

The measurement and calculation results when the measurement step length $d = 0.10$ m were close to the situation when $d = 0.05$ m, which can improve the Widess resolution while maintaining the peak characteristics. The measurement and calculation results under this condition are no longer listed separately.

In addition, the integer multiple decimation of the measurement data under the condition of $d = 0.05$ m can obtain part of the original measurement data under the condition of a larger measurement step. Numerical differential operation was performed on the extracted data, and its peak characteristics and Widess resolution were close to the case of $d = 0.05$ m. However, as the extraction multiple increased, the error between the measured value and the actual value of the conductor peak position gradually increased.

In summary, the experimental results are in good agreement with the simulation conclusions in Section 3.1; that is, there exists a local optimum value for the measurement step, and the selection of a suitable measurement step can reduce the error of conductor positioning.

4. Conclusions

An improved magnetic field method was proposed to locate unknown grounding grid conductors. The influence of the truncation error and rounding error of the magnetic field differential on the positioning of grounding grid conductors was studied, and the optimal measurement step size under different differential orders was given. By selecting the optimal measurement step size, the accuracy of the magnetic field differential method for locating the conductor of the grounding grid was improved. Through theoretical analysis and experimental verification, the following conclusions were drawn:

- A small measurement step causes a large rounding error in the numerical differentiation. With the increase of the measurement step, the truncation error caused by the numerical differentiation increases the error of the conductor positioning.
- The peak position deviation expectation and variance showed a trend of first decreasing and then increasing with the increase of the measurement step size. The measurement step size under different differential orders had a local optimum value, and the range of the local optimum value was given; this result is verified by simulation and experimental results.
- Choosing a reasonable measurement step size for different differential orders helps to improve the positioning accuracy of the grounding grid conductors.

The error analysis method in this paper expands the possibility of using the magnetic field differential method to reduce the error of grounding grid conductor positioning. At the same time, the results of this paper can provide a reference for the structural design of the magnetic field measurement sensor array, and the grounding grid conductor positioning instrument designed by the results of this study will have a higher conductor positioning accuracy, which is beneficial to the fault diagnosis work of large-scale substations.

Author Contributions: Conceptualization, F.Y. and Y.L.; methodology, S.L. and Y.L.; writing, S.L.; visualization, S.L. and S.W.; supervision, F.Y. and J.H. All authors have read and agreed to the published version of the manuscript.

Funding: This research received no external funding.

Institutional Review Board Statement: Not applicable.

Informed Consent Statement: Not applicable.

Data Availability Statement: Not applicable.

Conflicts of Interest: The authors declare no conflict of interest.

References

1. Liu, P.; Huang, C. Detecting Single-Phase-to-Ground Fault Event and Identifying Faulty Feeder in Neutral Ineffectively Grounded Distribution System. *IEEE Trans. Power Deliv.* **2018**, *33*, 2265–2273. [CrossRef]
2. Li, B.; Ren, X.; Li, B. Study on the Charge Transfer Criterion for the Pole-to-Ground Fault in DC Distribution Networks. *IEEE Access* **2019**, *7*, 102386–102396. [CrossRef]
3. Qiu, Y.; Thomas, S.; Fabijanic, D.; Barlow, A.J.; Fraser, H.L.; Birbilis, N. Microstructural Evolution, Electrochemical and Corrosion Properties of AlxCoCrFeNiTiy High Entropy Alloys. *Mater. Des.* **2019**, *170*, 107698. [CrossRef]
4. Feng, L.; Yan, A.; Meng, Y.; Hou, J. Investigation on Corrosion of Yttrium-Doped Magnesium-Based Sacrificial Anode in Ground Grid Protection. *J. Rare Earths* **2010**, *28*, 389–392. [CrossRef]
5. Liu, B.; Ma, H.; Xu, H.; Ju, P. Single-Phase-to-Ground Fault Detection with Distributed Parameters Analysis in Non-Direct Grounded Systems. *CSEE J. Power Energy Syst.* **2019**, *5*, 139–147. [CrossRef]
6. Fu, Z.; Wang, X.; Wang, Q.; Xu, X.; Fu, N.; Qin, S. Advances and Challenges of Corrosion and Topology Detection of Grounding Grid. *Appl. Sci.* **2019**, *9*, 2290. [CrossRef]
7. Xue, Y.; Chen, X.; Song, H.; Xu, B. Resonance Analysis and Faulty Feeder Identification of High-Impedance Faults in a Resonant Grounding System. *IEEE Trans. Power Deliv.* **2017**, *32*, 1545–1555. [CrossRef]
8. Qamar, A.; Shah, N.; Kaleem, Z.; Uddin, Z.; Orakzai, F.A. Breakpoint diagnosis of substation grounding grid using derivative method. *Prog. Electromagn. Res. M* **2017**, *57*, 73–80. [CrossRef]
9. Hu, J.; Hu, J.; Lan, D.; Ming, J.; Zhou, Y.; Li, Y. Corrosion Evaluation of the Grounding Grid in Transformer Substation Using Electrical Impedance Tomography Technology. In Proceedings of the IECON 2017—43rd Annual Conference of the IEEE Industrial Electronics Society, Beijing, China, 29 October–1 November 2017; pp. 5033–5038.
10. He, Y.; Shao, X.; Hu, J.; Liu, Y.; Jin, C.; Pan, J. Corrosion Condition Detect of Entire Grounding System in a 500 kV Converting Station Using Electrical Impedance Imaging Method. In Proceedings of the 2018 IEEE International Conference on High Voltage Engineering and Application (ICHVE), Athens, Greece, 10–13 September 2018; pp. 1–4.
11. Gomes, L.V.; de Macedo, E.C.T.; Albuquerque, T.C.; Guedes, E.C.; Junior, G.V.A.; de Castro, M.S.; Freire, R.C.S. Embedded System to Grounding Grid Diagnosis of Energized Substations. In Proceedings of the 2012 IEEE International Instrumentation and Measurement Technology Conference Proceedings, Graz, Austria, 13–16 May 2012; pp. 796–800.
12. Hu, H.; Luo, R.; Fang, M.; Zeng, S.; Hu, F. A New Optimization Design for Grounding Grid. *Int. J. Electr. Power Energy Syst.* **2019**, *108*, 61–71. [CrossRef]
13. El-Refaie, E.-S.M.; Elmasry, S.E.; Elrahman, M.K.A.; Abdo, M.H. Achievement of the Best Design for Unequally Spaced Grounding Grids. *Ain Shams Eng. J.* **2015**, *6*, 171–179. [CrossRef]
14. Dong, M.; Shi, Z.; Li, X.; Shao, G.; Yang, F.; Yao, D.; Zhang, K. A Diagnosis of Grounding Grid Corrosion Defects Based on Branch Voltage Disturbance. *IEEE Access* **2020**, *8*, 36749–36756. [CrossRef]
15. Chow, Y.; Salama, M. A Simplified Method for Calculating the Substation Grounding Grid Resistance. *IEEE Trans. Power Deliv.* **1994**, *9*, 736–742. [CrossRef]
16. Ma, J.; Dawalibi, F.P. Influence of Inductive Coupling between Leads on Ground Impedance Measurements Using the Fall-of-Potential Method. *IEEE Trans. Power Deliv.* **2001**, *16*, 739–743. [CrossRef]
17. Dawalibi, F. Electromagnetic Fields Generated by Overhead and Buried Short Conductors. Part 1 Single Conductor. *IEEE Power Eng. Rev.* **1986**, *1*, 112–119. [CrossRef]
18. Zhang, B.; Zhao, Z.; Cui, X.; Li, L. Diagnosis of Breaks in Substation's Grounding Grid by Using the Electromagnetic Method. *IEEE Trans. Magn.* **2002**, *38*, 473–476. [CrossRef]
19. Kai, L.; Fan, Y.; Songyang, Z.; Liwei, Z.; Jiayuan, H.; Xiaoyu, W.; Ullah, I. Research on Grounding Grids Imaging Reconstruction Based on Magnetic Detection Electrical Impedance Tomography. *IEEE Trans. Magn.* **2018**, *54*, 1–4. [CrossRef]
20. Song, H.; Dong, H.; Wang, X.; Tang, L. Non-Destructive Diagnosis of Grounding Grids Based on the Electromagnetic Induction Impedance Method. *Meas. Sci. Technol.* **2021**, *32*, 115901. [CrossRef]
21. Zhang, P.; He, J.-J.; Zhang, D.; Wu, L.-M. A Fault Diagnosis Method for Substation Grounding Grid Based on the Square-Wave Frequency Domain Model. *Metrol. Meas. Syst.* **2012**, *19*, 63–72. [CrossRef]
22. Ma, Y.; Karady, G.G. Investigating Grounding Grid Integrity Based on the Current Injection Method. In Proceedings of the 41st North American Power Symposium, Starkville, MS, USA, 4–6 October 2009; pp. 1–5.
23. Kostić, V.I.; Raičević, N.B. A Study on High-Voltage Substation Ground Grid Integrity Measurement. *Electr. Power Syst. Res.* **2016**, *131*, 31–40. [CrossRef]
24. Fan, Y.; Kai, L.; Liwei, Z.; Songyang, Z.; Jiayuan, H.; Xiaoyu, W.; Bin, G. A Derivative-Based Method for Buried Depth Detection of Metal Conductors. *IEEE Trans. Magn.* **2018**, *54*, 1–9. [CrossRef]
25. Kou, X.; Dong, M.; Yang, F.; Han, S.; Zhang, K.; Guo, L.; Ding, G. Design of Grounding Grid Conductor Positioning Device on the Magnetic Field Method. *Prog. Electromagn. Res. M* **2018**, *67*, 105–117. [CrossRef]
26. Arellano, Y.; Hunt, A.; Haas, O.; Ahmed, H.; Ma, L. Multiple Regression-Based Prediction Correlations for Enhanced Sensor Design of Magnetic Induction Tomography Systems. *Meas. Sci. Technol.* **2019**, *31*, 024002. [CrossRef]
27. Yu, C.; Fu, Z.; Wu, G.; Zhou, L.; Zhu, X.; Bao, M. Configuration Detection of Substation Grounding Grid Using Transient Electromagnetic Method. *IEEE Trans. Ind. Electron.* **2017**, *64*, 6475–6483. [CrossRef]

28. Wang, X.; Fu, Z.; Wang, Y.; Liu, R.; Chen, L. A Non-Destructive Testing Method for Fault Detection of Substation Grounding Grids. *Sensors* **2019**, *19*, 2046. [CrossRef]
29. Yu, C.; Fu, Z.; Hou, X.; Tai, H.-M.; Su, X. Break-Point Diagnosis of Grounding Grids Using Transient Electromagnetic Apparent Resistivity Imaging. *IEEE Trans. Power Deliv.* **2015**, *30*, 2485–2491. [CrossRef]
30. Wang, Y.; Huang, Y.; Zeng, X.; Wei, G.; Zhou, J.; Fang, T.; Chen, H. Faulty Feeder Detection of Single Phase-Earth Fault Using Grey Relation Degree in Resonant Grounding System. *IEEE Trans. Power Deliv.* **2017**, *32*, 55–61. [CrossRef]
31. Anggoro, B.; Yutadhia, R.E. The Grounding Impedance Characteristics of Grid Configuration. *Procedia Technol.* **2013**, *11*, 1156–1162. [CrossRef]
32. Bao, J.-F.; Li, C.; Shen, W.-P.; Yao, J.-C.; Guu, S.-M. Approximate Gauss–Newton Methods for Solving Underdetermined Nonlinear Least Squares Problems. *Appl. Numer. Math.* **2017**, *111*, 92–110. [CrossRef]
33. Gaudreau, P.; Hayami, K.; Aoki, Y.; Safouhi, H.; Konagaya, A. Improvements to the Cluster Newton Method for Underdetermined Inverse Problems. *J. Comput. Appl. Math.* **2015**, *283*, 122–141. [CrossRef]
34. Chunli, L.; Wei, H.; Degui, Y.; Fan, Y.; Xiaokuo, K.; Xiaoyu, W. Topological Measurement and Characterization of Substation Grounding Grids Based on Derivative Method. *Int. J. Electr. Power Energy Syst.* **2014**, *63*, 158–164. [CrossRef]
35. Qamar, A.; Umair, M.; Yang, F.; Uzair, M.; Kaleem, Z. Derivative Method Based Orientation Detection of Substation Grounding Grid. *Energies* **2018**, *11*, 1873. [CrossRef]
36. Qamar, A.; Uddin, Z.; Yang, F. Inverse Features Extraction for Substation Grounding Grid: Derivative and ICA Combinatorial Approach. *IET Gener. Transm. Distrib.* **2019**, *13*, 5457–5466. [CrossRef]
37. Xiaoyu, W.; Wei, H.; Fan, Y.; Liwei, Z.; Xingping, L. Topology Detection of Grounding Grids Based on Derivative Method. *Trans. China Electrotech. Soc.* **2015**, *7*, 73–78.
38. Qingyang, L.; Nengchao, W.; Dayi, Y. *Numerical Analysis*; Tsinghua University Press: Beijing, China, 2008.

sensors

MDPI

Article

Realization of a Test Tool for Diagnosis of Contact Resistance and Measurement of Selected Types of Conductive Materials

Petr Kacor [1,*], Petr Bernat [1], Tomas Mlcak [2] and Leopold Hrabovsky [3]

1 Department of Electrical Power Engineering at VSB, Technical University of Ostrava, 17. listopadu 2172/15, 70800 Ostrava-Poruba, Czech Republic; petr.bernat@vsb.cz
2 Department of Electrical Engineering at VSB, Technical University of Ostrava, 17. listopadu 2172/15, 70800 Ostrava-Poruba, Czech Republic; tomas.mlcak@vsb.cz
3 Department of Machine and Industrial Design at VSB, Technical University of Ostrava, 17. listopadu 2172/15, 70800 Ostrava-Poruba, Czech Republic; leopold.hrabovsky@vsb.cz
* Correspondence: petr.kacor@vsb.cz

Abstract: Contact connections in electrical machines and apparatus are important elements in the whole power supply network and a high level of reliability is expected there. Contact resistance is a fundamental criterion in the design of an electrical contact or contact system. The contact resistance should be as low as possible to minimize losses due to the current passage and the related heating of the contact connection. The value of the contact resistance depends on the material used, the value of the applied force, the type of contact, and, last but not least, the quality of the surface and chemical layers. In this paper, an initial diagnosis of the contact material is performed based on the determination of the sample's specific resistivity by the four-wire method and the evaluation of the measurement uncertainty. The work is followed by the design of a testing device that uses crossed bars to measure the change in contact resistance as a function of the magnitude of the applied force. An analysis of the sample mounting method is performed here using FEM simulations of the current field and shows the interaction between the holder and the sample in terms of current line transfer. The proposed system is then used for experimental measurements of the material-dependent coefficient K_C for verification of existing or newly developed materials in electrical engineering, where the values of the K_C coefficient are not known. Finally, the paper also deals with the measurement of fritting voltage for individual contact pairs having surface quality corresponding to brushing.

Keywords: electrical contact; specific resistivity; contact resistance; constriction resistance; fritting; diagnostic; crossed bars; measurement

Citation: Kacor, P.; Bernat, P.; Mlcak, T.; Hrabovsky, L. Realization of a Test Tool for Diagnosis of Contact Resistance and Measurement of Selected Types of Conductive Materials. *Sensors* **2023**, *23*, 5867. https://doi.org/10.3390/s23135867

Academic Editors: Michał Kunicki, Jan Fulneček and Paweł Rozga

Received: 1 June 2023
Revised: 20 June 2023
Accepted: 21 June 2023
Published: 24 June 2023

1. Introduction

Electrical apparatus plays a fundamental role in maintaining a reliable distribution of electrical energy and are an essential part of any transmission and distribution power network. The degree of reliability is greatly dependent on the reliability of the switching elements, which, among other things, provide the control and operation of the electrical power network and generally ensure the transport and distribution of the electric energy generated in the power plants to the consumers safely and reliably. Thus, the reliability of the electrical energy delivery and the reliability of the electrical apparatus are interrelated domains [1,2].

From the point of view of operational safety, the electrical apparatus must in general ensure galvanic insulation between the consumption and the supply. Fault situations such as overloads, overvoltage, or short circuits sometimes occur in networks and the electrical apparatus must also fulfill a safety and protection function [3,4].

In industrial applications, we also often find apparatus systems for switching or reconnecting transformer taps and voltage controls. In recent decades the trend is to replace these systems with power electronic equipment, although we still find applications with classical mechanical switching [5,6].

A shared feature of all the above switching structures is the contact system one of the most important parts of any switching apparatus [7]. In the contact system, there is a mutual movement of the current paths, and the switching on and off of the energy flow takes place. The interruption of the conduction path can be realized under different operating conditions, e.g., with current, without current, at different voltages, with high switching density, etc. [7,8]. In general, the requirements for contacts can be characterized as follows [9]:

- Low contacts resistance;
- High durability to mechanical wear (abrasion);
- High resistance to welding;
- High resistance to electric arc burning;
- Optimal influence on the development of deionization processes after extinguishing the switching arc (good switching capability).

A low contact resistance is an important factor in the choice of contact material. From the current conduction point of view, the contact material should have the highest possible conductivity. Such materials exist (Ag, Cu), but they are usually mechanically soft or sensitive to surface layers or the formation of welds [9,10]. Tungsten or molybdenum, for example, has a high withstand capability against arcing. However, low electrical conductivity is typical for these materials and they are not suitable for direct contact applications [10,11].

Fulfillment of all the above conditions and defined requirements for contacts are often contradictory. The problem of contact design is approached according to the application of the electrical apparatus [9,12]. Examples of typical contact shapes for electrical equipment are shown in Figure 1.

Figure 1. Examples of power apparatus contact systems: (**a**) Knife contact of fuse and fuse disconnector; (**b**) Low-voltage circuit breaker; (**c**) AC power motor contactor; (**d**) Old-style AC power motor contactor; (**e**) DC contactor without extinguishing chamber; (**f**) AC power switch.

With the development of electrical engineering, higher and higher requirements are placed on contact systems. In particular, reliability and a long lifetime are required. The description of electrical contacts implemented e.g., in busbar connections, switchgear systems, or electrical apparatus, and the search for interdependencies between materials, surface layers, surface quality, and wear, have been studied by many authors worldwide for several decades [10,12–19]. Currently, contact design and development use complex numerical models and simulations that analyze not only the mechanical (roughness, elasticity, plasticity) and electrical (conductivity, contact bridges, chemical layers) aspects of contacts but also deal with thermal aspects (heating, cooling, softening, and welding voltage). Even as the parameters needed to design a reliable contact system are continuously improved, we still cannot avoid validation measurements that verify the results of these simulations [20–24].

Measurement of contact resistance is necessary for the testing and development of new contact materials required for various working conditions and applications. The value of contact resistance is a crucial criterion in the design of a contact system. It should be minimal to reduce heat loss. The magnitude of the contact resistance is affected by a series of factors where the most important ones are the material properties (electrical conductivity), the existence of impurities, oxides, and chemical layers, the shape and type of contact (point, line, and surface) and the applied contact force (contact pressure) [9,10,24].

Online contact diagnostics can be performed after the unit is manufactured or while the unit is already operating in the substation. The overall condition of the switching apparatus, including its current paths and connecting terminals, is evaluated. The static method consists in injecting direct current in the switched-on state of the device and measuring the voltage drop, looking for changes in parameters during switching operations. [25–27]. The dynamic method injects a high value of direct current (more than 100 A) to ignite an electric arc. Measuring the waveform of the current and voltage during the switching of the device is the base of this method. The contact surface erosion as well as the shortening of arc contacts can be detected in time [28,29].

In general, offline diagnostics can be considered a situation where the measurement of diagnostic parameters is carried out on individual components of the device in a disassembled state. Using offline diagnostics, the quality of the contact connection can be determined, for example, by measuring the contact resistance on recently manufactured or already used contacts. The influence of switching frequency on the wear of the contact surfaces and the range of change of the contact resistance Rs can be determined in this way [12,20,30,31].

One of the possibilities to perform offline diagnosis is the evaluation of contact resistance, e.g., using crossed bars [9,10,12]. The advantage lies in capturing the ratio between the contact resistance R_S and the applied force F. With the knowledge of other parameters such as the surface hardness and the modulus of elasticity of the contact material, it is possible to determine, for example, the average resistivity of the surface layers. The importance of this type of measurement is evident in the development of new or optimized contact materials or metal alloys [32–35].

Contact properties are a rather complicated function and complex dependence of both physical properties (e.g., density, chemist, hardness, electrical/thermal conductivity, structure, etc.) and operating conditions (voltage, current, frequency, power factor, frequency, contact on/off velocity and force, etc.) [11,12]. Specialized techniques and equipment are required to evaluate these contact properties. In particular, contact resistance, opening, and closing velocity effect, contact bounce, contact welding measurement, erosion, etc. are evaluated. In contact resistance measurements, two approaches are used, namely, crossed-bar arrangement or gold probe measurements under very low mechanical and electrical load to avoid any mechanical or electrical damage to the surface layer [36–38].

The main motivation for the research activity was the need to create a measurement system that can verify the electrical parameters of newly developed material structures and alloys. These new structures were formed by combining highly electric conductive materials Al/Cu by rotary swaging, rolling, or extrusion. The cross-section of these materials can be of sandwich composition or the secondary material is embedded in the shape of segments, bars, or tubes. The manufacturing process conditions and the method of additional heat treatment are also different [39,40].

Such materials have good potential in electrical applications in the design of electrical machines (hybrid rotors of asynchronous motors) and apparatus (bus bars, input terminals, parts of conducting ways or contacts) [41–44]. The partial substitution of Cu by Al generally leads to a reduction in weight while improving the mechanical and performance properties of the solid conductor without significantly reducing its electrical conductivity.

Authors often cover the mechanical or thermal properties of newly developed structures in their studies, but the electrical parameters of these materials are rarely or only occasionally discussed [45,46]. This offers the opportunity to fill a gap in knowledge of the specific electrical parameters of Al/Cu structures and to use electrical resistivity measurements to determine important design parameters required in electrical engineering practice. As a secondary step, the properties of these materials concerning contact coupling can then be also validated.

In the reference literature, in most cases, only conceptual diagrams of the testing device or details focusing on the contact pair appear, but the overall view of the measuring system with its detailed description is missing and reproducibility is thus impossible [20,22,24,25]. Therefore, this paper aims to analyze and implement the design of a verification tool and in the first phase of the work to perform electrical conductivity measurements with an extension to the measurement of contact resistance parameters on selected materials.

Concerning the current state of the art, the paper presents a possible way to identify existing materials or newly developed alloys or structures. The obtained measurement outputs can be directly used in the design or optimization of contact systems of switching apparatus. The main objective and new contribution of this paper are then to investigate the properties of materials in the initial stage of production and the influence of subsequent heat treatment. The construction of the test system was designed, verified, and modified, and based on the comparison of experimental results of contact resistance with available theoretical results, limits and flaws of the device were found.

2. Materials and Methods

The analytical models for calculating the contact resistance are based on the fact that the current between two electrically conductive components passes through a defined metallic reduced contact surface (a-spot) with a specific geometry. The current that passes through this conductive contact spot causes a so-called constriction resistance Rc due to the increased current density. There are different models for the description of the constriction resistance of electrical contact, but the generally accepted model for the calculation of electrical contacts is the ellipsoidal model according to Holm [9–11]:

$$R_C = \frac{\rho}{2\pi \cdot a} tan^{-1}\left(\frac{\sqrt{\mu}}{a}\right) \tag{1}$$

where ρ ($\Omega \cdot$m) is the resistivity of the contact material and a (m) is the radius of the contact surface. The parameter μ defines the distance of the vertical semi-axis of the ellipsoidal potential surface from the contact surface. Considering the dimensions of the contact body in comparison to the dimensions of the contact surface, it can then be assumed that $\sqrt{\mu} \rightarrow \infty$. For the total constriction resistance of the two contacts, it can be written:

$$R_C = \frac{\rho}{2a} \tag{2}$$

Because metals are not clean, the passage of electric current can be affected by thin layers of oxides, sulfides, and other inorganic substances that are usually present on the contact surface of metals. As a result, the total contacts resistance R_S of the connection is the sum of the constricted resistance R_C and the film resistance R_f [10,12]:

$$R_S = R_C + R_f \tag{3}$$

The film resistance R_f can be calculated:

$$R_f = \frac{\sigma_f}{\pi \cdot a^2} \tag{4}$$

where σ_f ($\Omega \cdot$m^2) is the resistance per film area.

To determine the size of the contact area a (a-spot) at the contact of two elements, the deformation, and tension due to the applied force can be calculated for the purely elastic behavior of the material according to Hertz's theory. The assumption is the formation of a circular contact area [10,11]. The radius of the contact area for the shape of the sphere-plate contacts or the case of crossed cylindrical bars can be calculated as:

$$a = 1.11\sqrt[3]{\frac{F \cdot r}{E}} \tag{5}$$

where F (N) is the contact force, r (m) is the radius of the cross bars and E (Pa) is the tensile modulus of elasticity. In addition to the elasticity condition, the Formula (5) is derived with the assumption that there is no friction and the bodies have smooth and spherical surfaces.

If the contact force F is larger, a combination of elastic and plastic deformation or purely plastic deformation may occur at the location of contact. For practical purposes, it is appropriate to consider only either pure elastic or pure plastic deformation. The combination of both deformations in the design of contacts then leads to difficult calculations and considerations of an extensive character [10,11]. For the case of plastic deformation, the formula for the contact area radius a-spot of the form:

$$a = \sqrt{\frac{F}{\pi \cdot \sigma_{Pd}}} \tag{6}$$

where σ_{Pd} (Pa) is the compressive strength (contact hardness). By substituting the above Equations (5) and (6) into Equation (2) we obtain the final relations for the calculation of the constricted resistance for elastic contact (indexed as CE) and contact with plastic deformation (indexed as CP):

$$R_{CE} = \frac{\rho}{2.22 \cdot \sqrt[3]{\frac{r}{E}}} \cdot F^{-\frac{1}{3}} \tag{7}$$

$$R_{CP} = \frac{\rho}{2} \cdot \sqrt{\pi \cdot \sigma_{Pd}} \cdot F^{-\frac{1}{2}} \tag{8}$$

Thus, in the case of elastic deformation, the contact resistance is a function of the force with an exponent of $^1/_3$ and in the case of plastic deformation with an exponent of $^1/_2$. The choice of the correct relationship for calculating the contact resistance is difficult and for practical purposes should be chosen with care.

In the practical calculation of the contact resistance R_{CO}, the empirically derived relationship between contact resistance and contact force is preferably used [7,8]:

$$R_{CO} = K_C \cdot F^{-n} \tag{9}$$

The coefficient K_C generally depends on the contact material, the type of machining, and the condition of the contact surface and also includes the influence of surface layers. The exponent n is then chosen according to the type of contact (shape-dependent exponent of the contact force). Empirical determination of the contact resistance is also useful for flat or line contacts that cannot be easily calculated analytically [47–49].

Typical values of the material-dependent coefficient K_C and the contact force exponent n are given in Table 1; the label LC denotes Low Current contacts type [8,50].

Since the coefficient K_C includes the effect of surface layers and takes into account the character of the contact force, Equation (9) gives higher values of resistance with load in comparison with theoretical relationships. The advantage of the empirically derived coefficient K_C is that its value can be determined also for the combination of two different contact materials [8,9]. Table 1 focuses on power contact systems of electrical apparatus, where higher contact forces and plastic deformation are considered. For all types of contact interfaces, the force exponent n is always at least equal to $^1/_2$ or higher.

Table 1. Values of material-dependent coefficient K_C and shape exponent n.

Material	Babikov [8]	Kurbatov [50]	Shape Exponent	
	K_C (μΩ·kg^n)		Type of Contact	n (-)
Copper-Copper	80 ÷ 140	400	Flat—Flat	1
Alumin-Alumin	3000 ÷ 6700	3000 ÷ 6000	Sharp—Flat	0.5
Brass-Brass	670	670	Sphere—Sphere	0.5
Steel-Steel	7600	7600	Brush—Flat	1
LC Copper-Copper	-	90 ÷ 280	Bus Bar	0.5 ÷ 0.7

2.1. Design and Construction of the Measuring System

The basic concept of the contact resistance measurement system is based on the recommendations of Mr. Holm and uses a cross-bar arrangement of the sample. This arrangement is advantageous due to the expected shapes of the material samples to test, but mainly because it eliminates the influence of the solid body resistance of the bar [10]. Figure 2 shows the basic layout and conceptual CAD design.

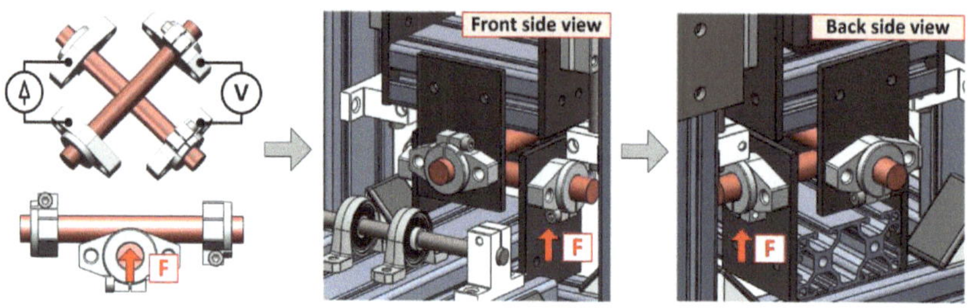

Figure 2. Basic design layout of testing tool with initial CAD design (The red arrow shows the direction of the applied force F).

The designed testing tool assumes the mounting of samples by holders which also operate as an electric current input and at the opposite side for voltage sensing. The insulating side plates to mount the contact holders are made of FR4 material. The measuring system is assembled from industrial aluminum profiles, allowing a large variation in shape. All construction elements are sufficiently rigid and the connections are made with bolts and tightening nuts.

The upper bar of the tested material is mounted in a holder on a vertically sliding base, on top of which a force gauge is inserted. The self-weight of the base and the force gauge is balanced by springs on both sides. The lower bar holder is part of a lever freely rotated in bearings around the shaft. A weight placed on the opposite side of the lever then generates the contact force, see Figure 3.

Figure 3. Detailed view of crossed bars and location of force gauge on testing tool (CAD design).

The inputs of the current clamps and the outputs of the voltage sensors are realized directly on the holders of the bars. The samples are mounted relatively freely in the holders so that they can be rotated during the contact resistance measurement and the "fresh contact surfaces" can be easily adjusted against each other. The complete realization of the measuring system with the stored weights and the detail of the brass crossing bars is shown in Figure 4a–c.

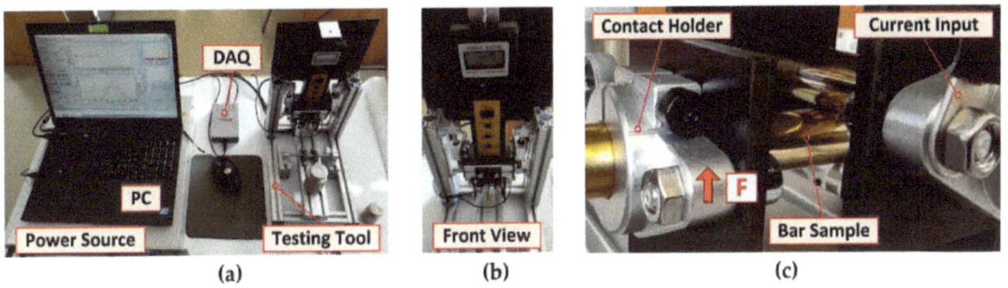

Figure 4. Assembled measuring system and detail of crossed bars in holders: (**a**) Overall view; (**b**) Front view; (**c**) Detail of crossed bars and sample holders.

The design of the measuring system assumed that there is no perfect contact between the holder and the measured sample over the entire circumferential surface. The supply current into the bar enters only from one side of the holder. This causes a deformation of the current lines in the bar sample, which is transferred from the holder space to the expected contact point (mid-length of the bar). The deformation of the current density decreases with distance from the holder, but we do not know at this time how strong and what effect the geometrical dimensions of the sample and its electrical conductivity have on this.

The requirement is that the deformation of the current lines due to the holder is not transferred to the point of contact. In other words, we were looking for the minimum necessary sample length L_{Smin}. FEM simulation of the current field distribution was used to solve this problem, see Figure 5.

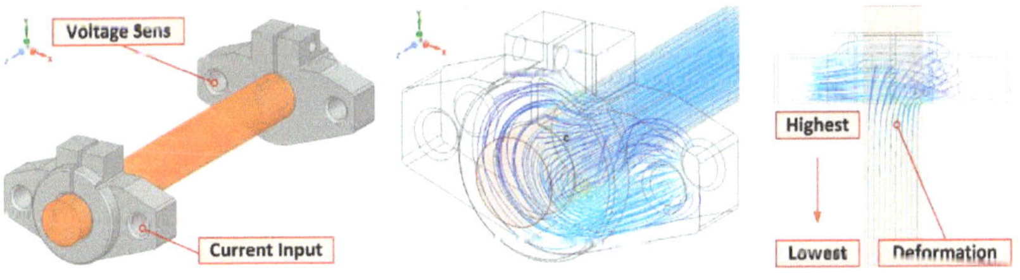

Figure 5. FEM simulation of current line deformation due to weak holder surface connection.

Figure 5 presents the FEM model of the holders and the attached bar. The plotted current lines show the extreme deformation at the point where the attached bar leaves the holder. This deformation of the current lines extends to the point of assumed contact with the second bar.

The degree of deformation of current lines is shown in the graph in Figure 6. Here the dependence of the current density J in the axial direction of the bar is plotted. The values are normalized with respect to the uniformity of the current density distribution at a very far point from the holder and expressed as a percentage. Three materials, copper, aluminum, and steel, were simulated. In the case of copper, the deformation of the current lines decreases at a much faster rate than that of the steel sample.

At a distance of L_X = 15 mm from the holder, the uniformity of the current density reaches about 99.9%. It can be considered homogeneously distributed over the entire cross-section. This length is then in fact equal to half of the minimum necessary distance between the holders. For safety, the real holder distance is doubled (L_{min} = 60 mm), as shown in Figure 6.

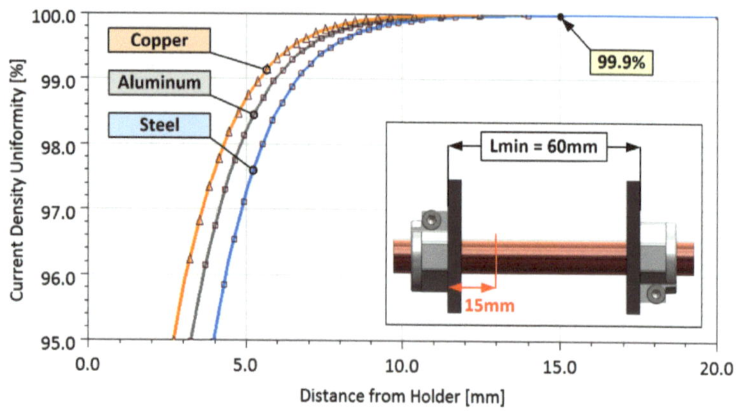

Figure 6. Degree of deformation of current lines with increasing distance from the holder.

The issue of current lines through different materials is discussed quite well in [11]. The deformation of the current lines is caused by the rapid change in geometry as well as the value of the electrical conductivity of the material, including the character of the contact between the holder and the sample. To complete the design of the sample holding system, Figure 7 shows the current line distribution solution in several considered situations.

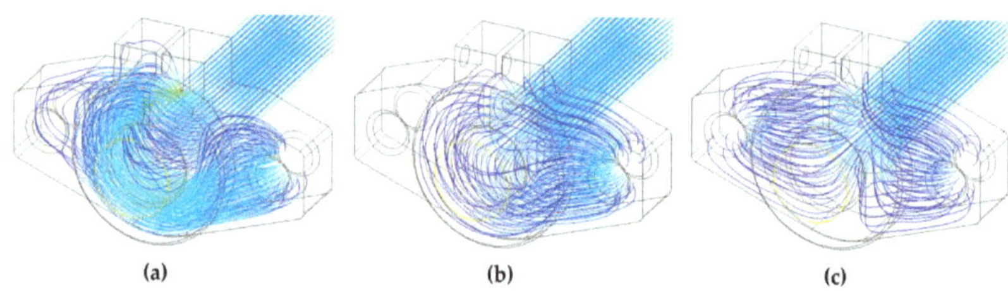

Figure 7. Selected simulations of current path entering the sample: (**a**) At the longitudinal edge of the holder only; (**b**) Across the entire contact area of the sample and the holder; (**c**) Across the entire contact area of the sample and the two-sided power supply of the holder.

Figure 7a shows a state where contact with the sample is ensured only at the longitudinal edge of the holder and simulates an insulating layer on the surface of the sample in contact with the holder. The other cases in Figure 7b,c then show ideal contact over the entire surface and current entering the holder from two sides. In the models, the interface layer between the holder and the sample was not considered, but the insulating region was simulated by inserting a non-conductive area without an air gap.

2.2. Samples Identification—Measurement of Electrical Resistivity

In the first step, the identification of the material samples was carried out. An important parameter from this point of view is the electrical resistivity ρ, which varies considerably for the considered samples of contact materials. Tabular data can be used for contact design [51], but a more accurate procedure is to measure the resistivity directly on the samples under study.

Ohm's law and the four-wire method [52] were used to determine the specific resistivity of the material samples. Each sample was supplied with DC using a current-limited EA-PS8080-120 power supply, and the voltage drop V between the electrodes connected to the sample surface was measured, see Figure 8. It was considered that the magnitude of the load current would generate only a negligible temperature rise. Due to the diameter of the sample D = 12 mm, the current value was set to I_{max} < 20 A.

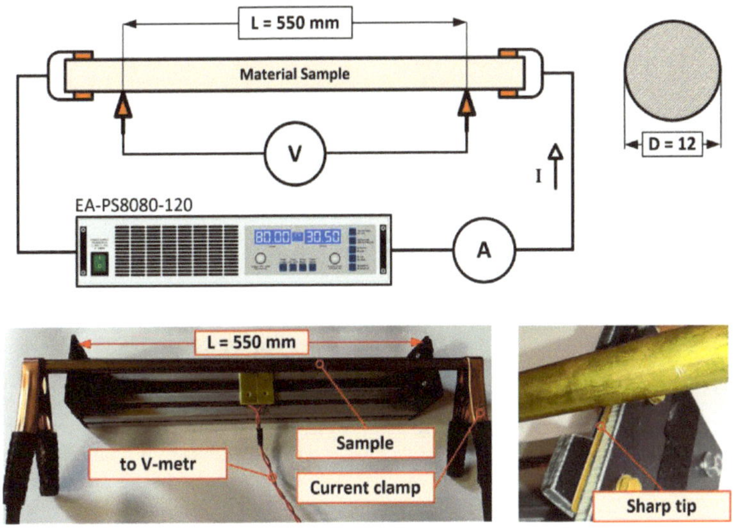

Figure 8. Schematic diagram of the circuit for measuring the resistivity and practical realization.

The magnitude of the load current I when measuring resistance R is always a compromise between the voltage drop V on the voltage electrodes and the overall temperature rise of the sample $\Delta \upsilon$. The value of current was chosen considering adiabatic heating without dissipation of heat to the surroundings. The total sample temperature rise over time can be determined from the:

$$Q = m \cdot c_m \cdot \Delta\vartheta = R \cdot I^2 \cdot t \tag{10}$$

where Q (J) is the heat, m (kg) is the sample weight, c_m (J·kg^{-1}·K^{-1}) is the thermal capacity, $\Delta\upsilon$ (K) is the temperature rise, R (Ω) is the sample resistance, I (A) is the magnitude of loading current, and t (s) is the time of its passing. After the modifications of (9), the temperature rise of the sample can be calculated:

$$\Delta\vartheta = \frac{R \cdot I^2}{m \cdot c_m} \cdot t = \frac{\rho \cdot I^2}{A^2 \cdot c_V} \cdot t \tag{11}$$

where ρ (Ω·m) is the specific electrical resistivity, A (m^2) is the sample cross-section and c_V (J·m^{-3}·K^{-1}) is volumetric thermal capacity. At a constant value of load current, I = 20 A, the sample with the lowest electrical conductivity will achieve the highest temperature rise. Therefore, we add the values of conductivity corresponding to 15% IACS (similar to the steel S235JR sample):

$$\Delta\vartheta = \frac{\rho \cdot I^2}{A^2 \cdot c_V} \cdot t = \frac{115 \cdot 10^{-9} \cdot 20^2}{\left(\frac{\pi \cdot 0.012^2}{4}\right)^2 \cdot 3.69 \cdot 10^6} \cdot 1 = 0.001\ \mathrm{K} \tag{12}$$

The result of (12) shows that for the sample with the lowest conductivity (S235JR) when a current of I = 20 A is applied, the temperature increases by 0.001 K every second. The time for one measurement and reading of the electrical quantities after the current was switched on never exceeded t_M = 10 s. This was followed by a break of approximately t_B = 120 s to cool down.

Equation (13) was used to calculate the specific electrical resistance and Equation (14) was derived by adding the geometric dimensions of the measured sample:

$$R = \rho \cdot \frac{L}{A} \tag{13}$$

$$\rho = R \cdot \frac{A}{L} = \frac{V \cdot A}{I \cdot L} = \frac{\pi}{4} \cdot \frac{V \cdot D^2}{I \cdot L} \tag{14}$$

where ρ ($\Omega \cdot$m) is the specific electrical resistivity, A (m^2) is the cross-section of the sample, V (V) is the voltage drop, D (m) is the diameter of the sample, I (A) is the DC supply current and L (m) is the spacing between voltage electrodes located on the sample surface.

As this is generally a task of determining the material resistivity ρ using indirect measurements, it was necessary to quantify the measurement uncertainties Type A and B for all measured data and then integrate these influences into the final result [53,54]. An example of the evaluation is performed on a brass bar with the label CuZn39Pb3.

2.2.1. Direct Measurement of the Diameter and the Spacing of the Sensing Electrodes

The geometric dimensions of the sample were measured using two Mitutoyo digital calipers. One was used to measure the sample diameter D and the other to measure the electrode spacing L. A total of ten measurements (n = 10) were completed under the same conditions in both cases. The bar diameter D was measured at randomly selected points, namely at both ends and in the middle of the sample length.

The measured data were tested for normality using the Shapiro-Wilk test and at a significant level of α = 0.05, the hypothesis that the measured statistical sample is normal was confirmed [55]. The estimated value of the final brass bar diameter D is given by the arithmetic mean of the individual measurements D_k:

$$\overline{D} = \frac{1}{n} \sum_{k=1}^{n} D_{,k} \tag{15}$$

The corresponding uncertainty of Type-A associated with the estimation of D is determined as the experimental standard deviation of the average. For the case of repeated measurements of $n \geq 10$ then:

$$u_A(\overline{D}) = \sqrt{\frac{1}{n(n-1)} \sum_{k=1}^{n} \left(D_{,k} - \overline{D}\right)^2} \tag{16}$$

Equations (15) and (16) are also used for the case of measuring the spacing L of the voltage electrodes. The processed data are shown in Table 2.

Table 2. Measured and evaluated data for uncertainty Type A for D and L.

n	Diameter D (mm)	Length L (mm)
1	11.97	550.13
2	12.03	550.39
3	11.91	550.41
4	11.94	550.03
5	12.05	549.95
6	12.00	549.93
7	11.93	550.11
8	11.95	550.31
9	12.00	550.27
10	12.02	550.21
Arithmetic average	11.980	550.174
u_A (D), u_A (L)	0.0148	0.0546

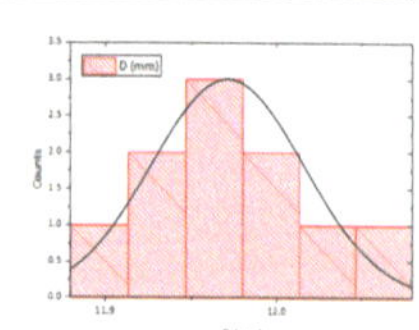

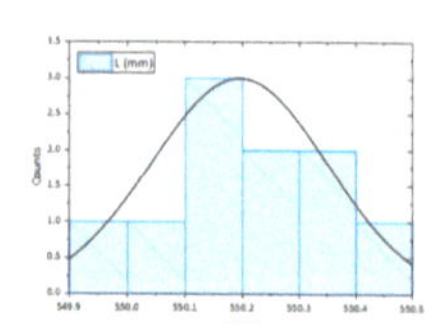

A certificate is available from the caliper manufacturer that indicates that the instrument has a resolution error of δ_1 = (0.02 + 0.00005$\cdot L$) within the length measurement interval L = (0 ÷ 150) mm. This gives a total resolution error of δ_1 = 0.0206 mm when taking into account the diameter of D = 12 mm. In Type B uncertainty we also consider operator influences representing the imperfection of the alignment of the measuring instrument for the diameter of the bar, varying pressure of the measuring plates, etc. We consider the magnitude of this error to be approximately equal to the resolution error of the instrument, i.e., $\delta_2 = 0.021$ mm. For both errors δ_1 and δ_2, we also assume a uniform rectangular distribution under, then [53]:

$$u_{B1}(\overline{D}) = \frac{\delta_1}{\sqrt{3}} \qquad u_{B2}(\overline{D}) = \frac{\delta_2}{\sqrt{3}} \tag{17}$$

The final B-Type standard uncertainty of the bar diameter estimation D is calculated as a summarization of both uncertainties:

$$u_B(\overline{D}) = \sqrt{u_{B1}{}^2(\overline{D}) + u_{B2}{}^2(\overline{D})} \tag{18}$$

The combined uncertainty is finally obtained by adding $u_A(D)$ and $u_B(D)$ and corresponds to the relation:

$$u_C(\overline{D}) = \sqrt{u_A{}^2(\overline{D}) + u_B{}^2(\overline{D})} \tag{19}$$

The relations (16) to (19) are also used to process the values measurements of the electrode spacing L. The processed data for diameter D and electrode spacing L are shown in Table 3.

Table 3. Evaluated data for combined uncertainty and coverage factor $k = 2$ for D and L.

Value	Estimation (mm)	Standard Uncertainty (mm)	Distribution	Coefficient of Sensitivity	Contribution (mm)
D	11.980	0.015	normal	1	0.015
Scale δ_1 (D)	-	0.012	rectangle	1	0.012
OP infl δ_2 (D)	-	0.012	rectangle	1	0.012
D	11.98	-	-	-	0.023
L	550.174	0.055	normal	1	0.055
Scale δ_1 (L)	-	0.061	rectangle	1	0.061
OP infl δ_2 (L)	-	0.060	rectangle	1	0.289
L	550.174	-	-	-	0.299
Bar Diameter D (mm):		11.98 ± 0.05			$k = 2$
Electrode Spacing L (mm):		550.17 ± 0.60			$k = 2$

2.2.2. Direct Measurement of Current I and Voltage Drop V

The procedure for the evaluation of the measured current and voltage is the same as for the determination of the Type A and B uncertainties in Section 2.2.1. For both electrical quantities, the Shapiro-Wilk test was performed, the arithmetic average of 10 sample measurements under the same conditions was determined (15) and the standard deviation was calculated according to (16). The processed data are presented in Table 4.

The current I passed through the measured sample was measured with a digital ammeter. TRMS multi-meter, Model ANENG 870 (as the ammeter) with 20,000 counts LCD. The accuracy of this multi-meter for the DC range was ±(0.5% reading + 3 digits) with a resolution of 0.001 A. The voltage V at the sensing electrodes was measured with a digital voltmeter. TRMS multi-meter, Model ANENG 870 (as the voltmeter) with 20,000 counts LCD and input impedance > 10 MOhm.

Table 4. Measured and evaluated data for uncertainty Type A for *I* and *V*.

n	Voltage *V* (mV)	Current *I* (A)
1	6.401	19.139
2	6.396	19.141
3	6.361	19.135
4	6.440	19.144
5	6.414	19.132
6	6.385	19.140
7	6.366	19.138
8	6.389	19.143
9	6.369	19.137
10	6.407	19.133
Arithmetic average	6.393	19.138
u_A (*V*), u_A (*I*)	0.0077	0.0013

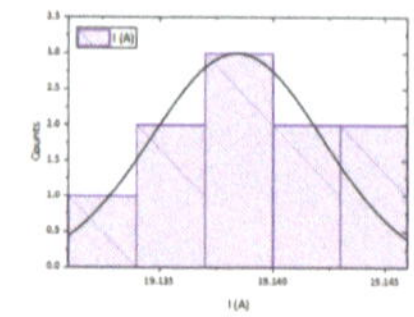

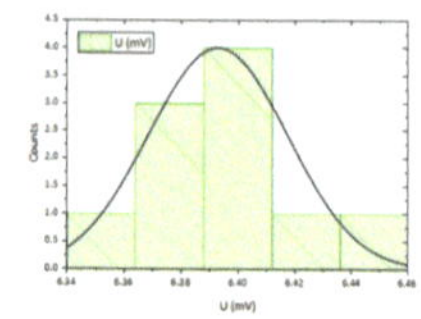

The accuracy of this multi-meter for the DC millivoltage range was ±(0.05% reading + 3 digits) with a resolution of 0.001 mV. The following relationship was used to calculate the measurement uncertainty of the digital meter:

$$u_B(\overline{V},\overline{I}) = \pm\left(\frac{MV}{100}\cdot|\delta_{RDG}| + DIGS\cdot RES\right) \tag{20}$$

where *MV* is the measured value, *DIGS* is the number of digits and *RES* is the resolution of multi-meter at selected range. The processed data are presented in Table 5.

Table 5. Evaluated data for combined uncertainty and coverage factor *k* = 2 for *I* and *V*.

Value	Estimation (A, mV)	Standard Uncertainty (A, mV)	Distribution	Coefficient of Sensitivity	Contribution (A, mV)
I	19.138	0.0013	normal	1	0.0013
A-meter δ_1 (*I*)	-	0.0569	rectangle	1	0.0569
I	19.14	-	-	-	0.057
V	6.393	0.0077	normal	1	0.0077
V-meter δ_1 (m*V*)	-	0.0036	rectangle	1	0.0036
V	6.39	-	-	-	0.0085
Current *I* (A):		19.14 ± 0.11			*k* = 2
Voltage *V* (mV):		6.39 ± 0.02			*k* = 2

The specific electrical resistivity ρ of the measured sample is calculated using the derived relation (14) by adding the values from Tables 3 and 5:

$$\rho = \frac{\pi}{4}\cdot\frac{V\cdot D^2}{I\cdot L} = \frac{\pi}{4}\cdot\frac{6.39\cdot 10^{-3}\cdot\left(11.98\cdot 10^{-3}\right)^2}{19.14\cdot 550.14\cdot 10^{-3}} = 68.41\cdot 10^{-9}\ \Omega\cdot\text{m} \tag{21}$$

As the resistivity of the sample is determined by indirect measurement, it is necessary to combine all the estimated uncertainties in the current, voltage, and geometric dimensions of a sample by applying the law of uncertainty propagation to the measurement model used [53,54]. We did not consider any covariance and the measured values of the voltmeter, ammeter, and calipers were not correlated.

The task was to determine the individual sensitivity coefficients Ai using partial derivatives of all variables:

$$\frac{\partial\rho}{\partial D} = \frac{\pi}{2}\cdot\frac{V\cdot D}{I\cdot L} \qquad \frac{\partial\rho}{\partial L} = -\frac{\pi}{4}\cdot\frac{V\cdot D^2}{I\cdot L^2} \qquad \frac{\partial\rho}{\partial I} = -\frac{\pi}{4}\cdot\frac{V\cdot D^2}{I^2\cdot L} \qquad \frac{\partial\rho}{\partial V} = \frac{\pi}{4}\cdot\frac{D^2}{I\cdot L} \tag{22}$$

The final standard uncertainty of the specific resistance of the bar sample is given by combining all these uncertainties as follows:

$$u_\rho = \sqrt{\left(\frac{\partial \rho}{\partial D}\right)^2 \cdot u_c^2(D) + \left(\frac{\partial \rho}{\partial L}\right)^2 \cdot u_c^2(L) + \left(\frac{\partial \rho}{\partial I}\right)^2 \cdot u_c^2(I) + \left(\frac{\partial \rho}{\partial V}\right)^2 \cdot u_c^2(V)} \quad (23)$$

The processed data are shown in Table 6.

Table 6. Evaluated data for uncertainty of specific resistivity of sample with coverage factor $k = 2$.

Value	Estimation	Standard Uncertainty	Sensitivity Coefficient	Uncertainty Contribution
D	$11.98 \cdot 10^{-3}$ m	$22.2 \cdot 10^{-6}$ m	$11 \cdot 10^{-6}$ Ω	$244 \cdot 10^{-12}$
L	$550.17 \cdot 10^{-3}$ m	$300 \cdot 10^{-6}$ m	$-124 \cdot 10^{-9}$ Ω	$-37 \cdot 10^{-12}$
I	19.14 A	0.057 A	$-3.6 \cdot 10^{-9}$ Ω·m/A	$-205 \cdot 10^{-12}$
V	$6.39 \cdot 10^{-3}$ V	$8.5 \cdot 10^{-6}$ V	$11 \cdot 10^{-6}$ m/A	$94 \cdot 10^{-12}$
ρ	$68.41 \cdot 10^{-9}$	-	-	$334 \cdot 10^{-12}$
Resistivity ρ (nΩ·m):		68.41 ± 0.67		$k = 2$

The remaining material samples, ETC copper, aluminum, and steel bar, were measured in a similar procedure. The summary results of the specific electrical resistivity measurements are shown in Table 7. This table also includes the modulus of elasticity values collected from the material supplier's data sheets. For comparison, tabulated values of specific resistivity obtained from reference literature are added here [11,51].

Table 7. Basic material parameters of samples for contact resistance measurements.

Material	Name	Resistivity ρ (nΩ·m)	Reference ρ_D (nΩ·m)	Modulus of Elasticity E [GPa]
Copper	Cu-ETC	17.57 ± 0.24	$16.5 \div 18$	117 [56]
Aluminum	AlCu4PbMg	47.91 ± 0.51	$45 \div 49$	74 [57]
Brass	CuZn39Pb3	68.41 ± 0.67	$67 \div 58$	96 [58]
Steel	S235JR	149.70 ± 1.40	$140 \div 150$	210 [59]

2.3. Preparing Samples for Contact Resistance Measurements

All samples were cut to the required length once the resistivity measurements had been completed. Samples with heavy surface contamination were degreased. The surface was sanded first with coarse and then fine sandpaper. The final mechanical surface treatment was brushing. Immediately after the surface treatment, a layer of technical lubricant was applied to the samples to protect their surface from external influences, see Figure 9.

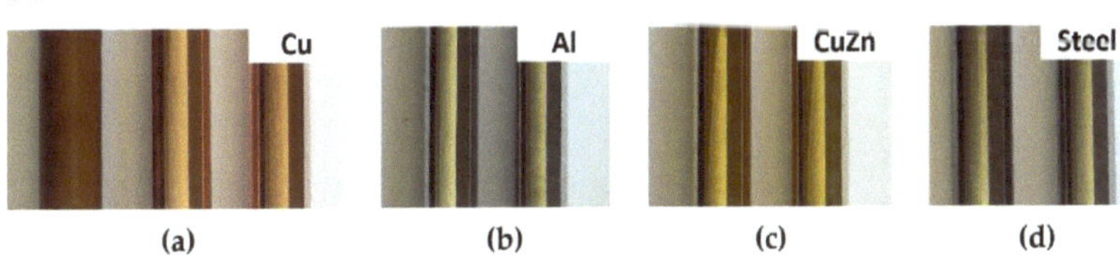

Figure 9. Example of the surfaces of samples: (**a**) Copper; (**b**) Aluminum; (**c**) Brass; (**d**) Steel.

The difference in surface quality between the original copper bar sample and the sample after brushing is shown in Figure 9a. The surfaces of the aluminum, brass, and steel bar samples were prepared in the same way, see Figure 8b–d. The sample surfaces have much less roughness and contamination compared to the contact surfaces commonly found in electrical power apparatus, see Figures 1 and 9.

2.4. Procedure for Measuring the Contact Resistance

The measured sample material was mounted in the holders. The contact force F was set using a weight inserted on the arm of the device and measured with a force gauge. The verification of the contact force was carried out without an electric current. A power supply was then connected with the current limited to $I = 1$ A and the magnitude of the voltage drop V across the contacts was measured simultaneously. After measuring the voltage drop, the power supply was briefly switched off and then on again. The measurements were carried out 10 times in this way. After this cycle, the contacts were unloaded and the sample material in both holders was rotated. After the contact force was applied again, the measurement cycle was repeated.

A total of 5 contact positions for the same magnitude of contact force F were measured using this sequence. Supply current I and voltage V were measured using a DAQ card and LabView application with continuous recording of measured values. After the measuring cycle, the contact force F was changed and the procedure was repeated in the total range of loading $F = \{0.1, 0.5, 1, 2, 5, 10\}$ N.

The waveform of the captured voltage drop V and current I during the measurement for a contact force $F = 5$ N on the brass bars is shown in Figure 10. All the material samples used were measured using the same procedure.

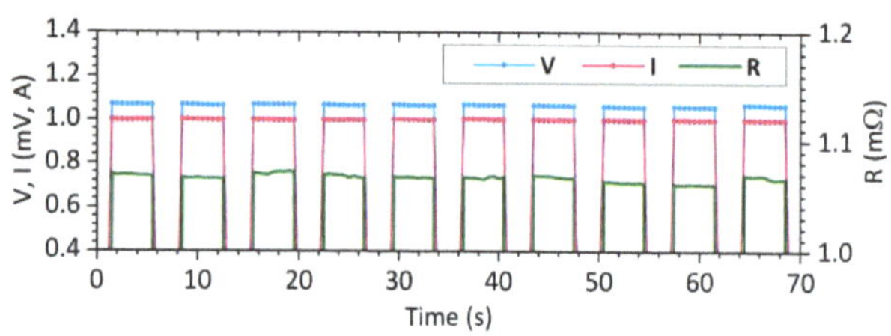

Figure 10. Time dependence of current, voltage and resistance during measurement, $F = 5$ N.

3. Results

The measured data were processed for each material and the results of the contact resistance plotted on a graph $R_{CO} = K_C F^{-n}$ in logarithmic scale. The values of the coefficient Kc and exponent n for all contact material combinations are shown in Table 8.

Table 8. Values of material-dependent coefficient K_C and shape-exponent n.

Material Combination	HOLM (n = 1/3)	Experimental Values	
	K_{TE} (mΩ·N^n)	K_{EX} (mΩ·N^n)	n (-)
Copper-Copper	0.213	0.51	0.43
Alumin-Alumin	0.499	4.93	0.40
Brass-Brass	0.776	1.95	0.39
Steel-Steel	2.344	13.4	0.47

The dependence of the contact resistance R_{SCu} on the applied contact force F for crossed copper bars is shown in Figure 11.

The black dashed line shows the dependence under pure elastic deformation of the contact point using (7) with values of the resistivity and modulus of elasticity parameters from Table 7. The red dashed lines indicate the range of contact resistance extracted from Table 1, where a full plastic deformation of the contact area is assumed. The exponent n here equals $-^1/_2$ and dependencies were derived from measurements on real contacts of power electrical connections.

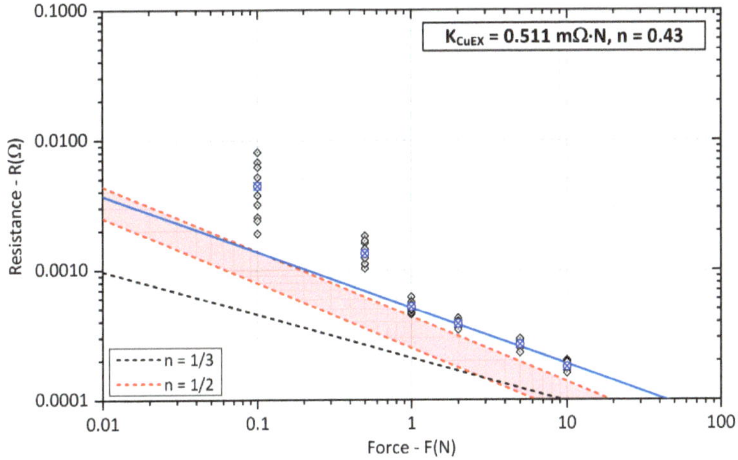

Figure 11. Experimental evaluation of the K_{Cu} coefficient and exponent n for copper bars (The blue cross represents the average value of the measured points).

Finally, the blue line corresponds to the experimentally determined parameters of the coefficient K_{CuEX} with associated exponent n. For lower contact forces $F = 0.1$ N and 0.5 N, there was a significant deviation in the measured values and these results were not included in the overall result. Measurements at low contact forces $F < 1$ N were also very sensitive to small vibrations and shaking, we decided to exclude them from all evaluations. Thus, the blue line was obtained by interpolating the average values of the experimentally measured resistances in the range of $F = \{1, 2, 5, 10\}$ N.

Contact resistance R_{SAl} measurements on aluminum crossed bars were relatively difficult to perform, the results of which are shown in Figure 12. The experimentally determined coefficient K_{AlEX} reaches lower values when compared to the data in Table 1, the exponent n adjusts the slope of the line more to the elastic form of deformation. The blue line was obtained by interpolating the average values of the experimentally measured resistances in the range of $F = \{1, 2, 5, 10\}$ N.

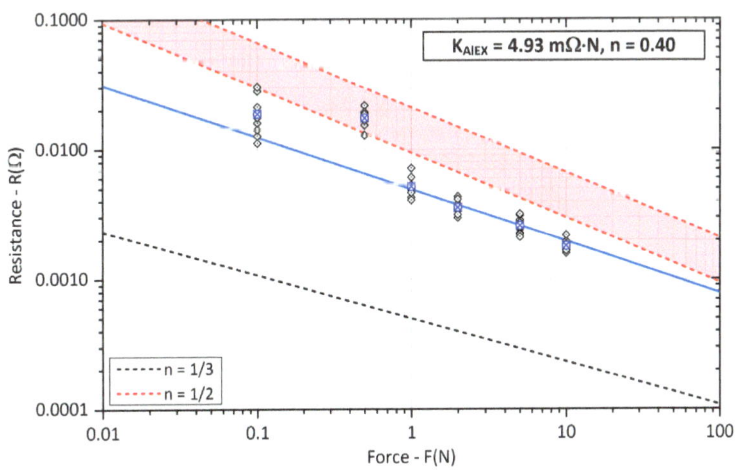

Figure 12. Experimental evaluation of the K_{Al} coefficient and exponent n for aluminum bars (The blue cross represents the average value of the measured points).

For small contact forces $F < 1$ N, there was also a significant variance in the measured resistance values. In some measurements, the contact resistance even showed a value of $R_C \to \infty$, although the force gauge indicated a load of $F = 0.5$ N. This is probably due to the very hard oxides Al_2O_3 on the surface of the aluminum bars. Treatment with technical grease immediately after finishing the contact surface did not help either. When the measured bars were touched, there was no friction of the contact surfaces, but only direct contact, thus the removal of the surface layers was ineffective and fully affected the contact resistance value.

The results of the contact resistance R_{SCuZn} measurements on brass-crossed bars are shown in Figure 13. Compared to the measurements on aluminum bars, the experimentally determined coefficient K_{CuZnEX} is smaller, although the electrical conductivity of brass is much lower. This is confirmed by the values obtained from the literature reported in [8] and captured in Table 1. Here, the K_{CuZnEX} coefficient reaches approximately 2.5 times lower values than K_{AlEX} for aluminum, and the surface layers break down significantly better.

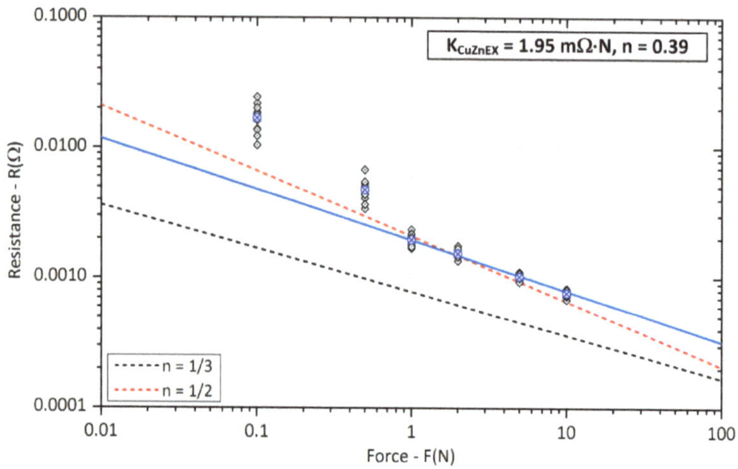

Figure 13. Experimental evaluation of the K_{CuZn} coefficient and exponent n for brass bars (The blue cross represents the average value of the measured points).

The measurements of contact resistance R_{SSteel} on steel crossed bars are shown in Figure 14. Compared to the other types of measured samples, this is a very hard material with the lowest electrical conductivity (12% IACS). The value of coefficient $K_{SteelEX}$ shows better results than Table 1. The value of the exponent n is the closest to the plastic deformation of the contact area of all tested contact materials.

Measurement of Fritting Voltage and Change of Contact Resistance with Current Loading

In the next step of the measurement, the contact materials were loaded with a higher current value, which reached the magnitude of $I_{max} = 120$ A. The waveform and the overall evolution of the voltage drop and the change of the contact resistance R_S after load removal were monitored.

The total measurement time was in the range of $t = 15$ s. There was an increasing current in the first half and a decrease in the second half of the interval. Figure 15 shows typical current and voltage waveforms recorded during the experiment on the crossed copper bars. The contact force was chosen at a lower value, a little below $F = 1$ N. Figure 15 also shows a V-I plot highlighting the change in contact resistance before and after the maximum load current is reached and showing the fritting at the contact location. The voltage at the contacts rises rapidly at first. As the current increases, the temperature of the contact spot becomes extremely high and the voltage drop grows.

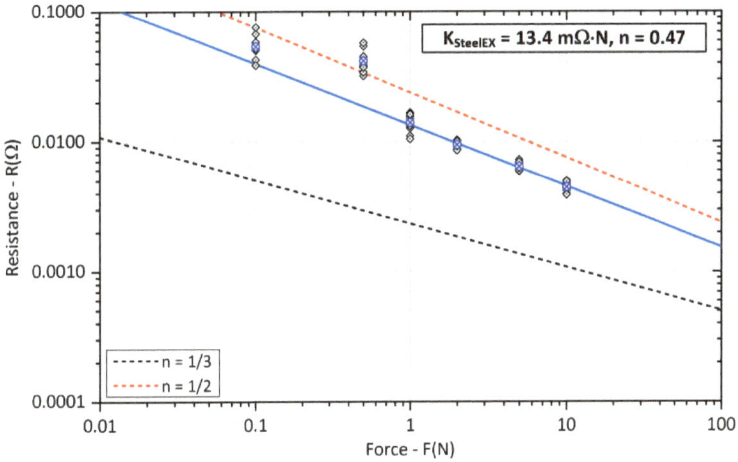

Figure 14. Experimental evaluation of the K_{Steel} coefficient and exponent n for steel bars (The blue cross represents the average value of the measured points).

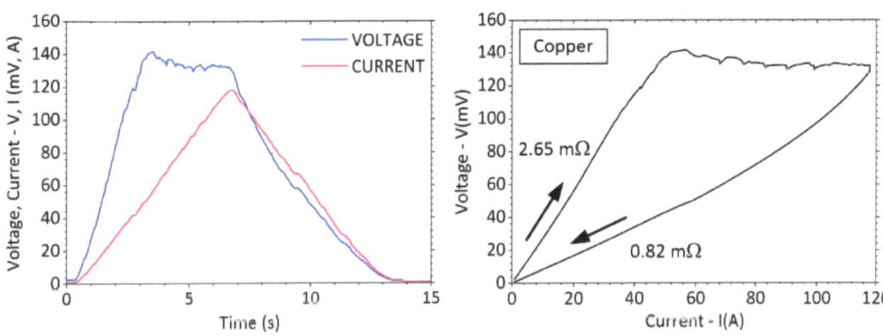

Figure 15. Current and voltage waveform with V-I characteristic for copper bars.

A significant break in the voltage waveform appears at V = 140 mV. When the current is decreased, the voltage then falls along a different trajectory with a lower slope. Proportionally, the value of the contact resistance is reduced by about 2.65/0.82 = 3.2 times. No welding of the contact surfaces occurred after the weights were removed.

The fritting voltage for aluminum bars reaches a higher value than for copper, approximately V = 245 mV, see Figure 16. Around a current of I = 90 A, there is a small drop in voltage and probably a final growth of the contact area.

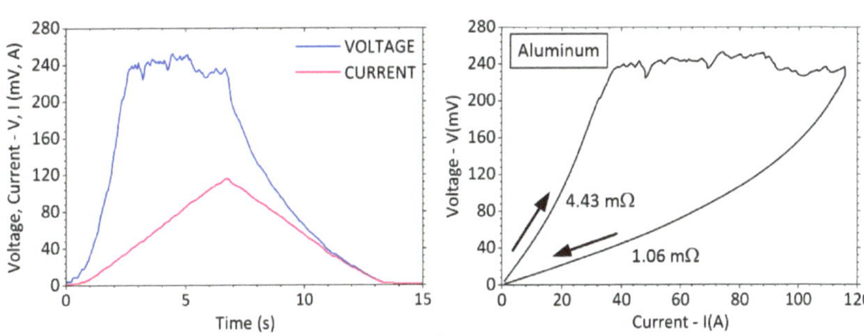

Figure 16. Current and voltage waveform with V-I characteristic for aluminum bars.

After the load current is reduced, the voltage falls along a lower curve to a final value of 1.06 mΩ. Proportionally, the value of the contact resistance decreased by about 4.43/1.06 = 4.2 times. No signs of welding were observed after the weights were removed and the contact surfaces were uncoupled.

During the measurement of brass bars, practically all experiments showed a slight drop in the voltage curve in the range of V_1 = 100 mV, but more often at V_2 = 135 mV. The voltage continued to increase with a slightly lower slope and the final fritting voltage value was then V_F = 220 mV. Occasionally a small peak occurred almost reaching a magnitude of V_P = 240 mV. A typical waveform of this can be seen in Figure 17.

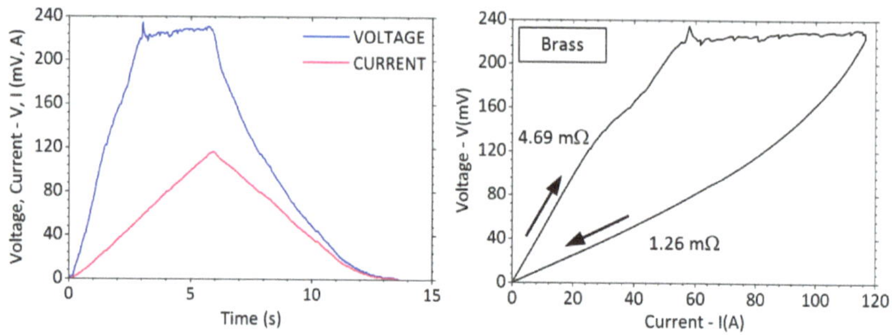

Figure 17. Current and voltage waveform with *V-I* characteristic for brass bars.

When the current was further raised, the voltage on the contacts increased only very slowly. After the load current was reduced, the voltage decreased with a lower slope to the final value of the contact resistance of 1.26 mΩ. Proportionally, the value of the contact resistance decreased about 4.69/1.26 = 3.7 times. After the weights were removed and the contact surfaces were uncoupled, there was no sign of welding.

In the measurements of the steel cross bars, the voltage on the contacts increased extremely fast already at 10% of the maximum current (I_1 = 12 A), see Figure 18. At a current of approx. I_2 = 80 A there was a violent rise in voltage and subsequent spikes reaching up to V_P = 380 mV in the peak. The voltage spikes continued until the maximum load current was reached without stabilization. The voltage spikes were also characterized by a significant sound effect.

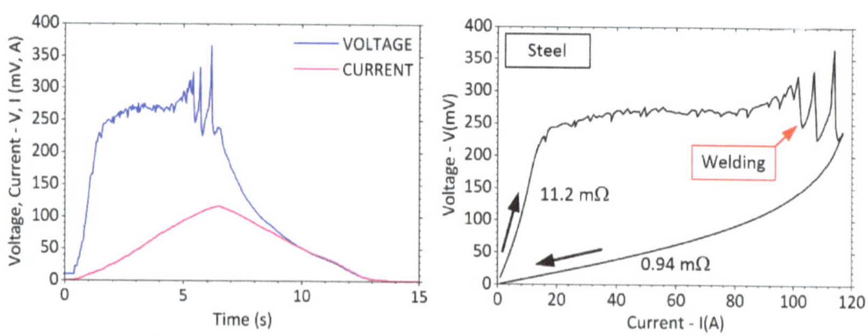

Figure 18. Current and voltage waveform with *V-I* characteristic for steel bars.

After the load current was switched off and the weights were unloaded, a significant force had to be applied to separate the contact surfaces. From this, it was considered that the welding voltage was achieved in the experiment. The *V-I* graph shows a significant change in the slope of the voltage rise. Proportionally, the value of the contact resistance decreased about 11.2/0.97 = 12 times.

4. Discussion

Due to the lack of more comprehensive information regarding the design of the contact loading system, we took our approach and designed a simple lever system according to Dr. Holm's schematics. The loading of the contacts is realized by calibrated weights and the force is measured by a force gauge. As simple as the proposed design is, the overall structural stiffness of the system is not optimal.

The design of the bar holders appears to be problematic, where applied forces $F > 5$ N lead to the bending of the specimen and a reduction in the contact force. For smaller forces $F < 1$ N, on the other hand, the influence of the connection and elasticity of the supply wires causes very poor repeatability of the experiments. This is reflected, among other things, by large deviations of the measured values in the case of smaller applied forces and is evident in all tested samples. The difference between the average and the two extreme minimum and maximum values reaches, for example, ±140% for copper bars and a load force of $F = 0.5$ N and ±200% for a force of $F = 0.1$ N. In the case of aluminum bars, this range of minimum and maximum resistance is smaller, but the contact resistance remains almost constant at both applied forces $F = 0.1$ N and 0.5 N. The above, due to the design limitations, was the main reason for the exclusion of contact resistance values at low values of the load force. The coefficient K_C was then determined from the contact resistance values at load $F > 1$ N.

The use of a frame structure with a crossbeam and vertically acting force, e.g., through a pneumatic cylinder, seems to be generally better than the originally designed system with an aluminum rotating arm and weight. The frame-enclosed construction with a crossbeam (a design well-known from pressing machines) provides better utilization of the applied force at the contact point. Force measurement can be realized by a strain gauge, which can be located under the holder at the bottom of the support. In the case of higher forces, this solution offers compensation for frame deformation, as the strain gauge captures the total resultant force. The expected improvement when using a frame construction is a higher stability of the applied force and a reduction in vibrations and shaking caused by the environment.

Furthermore, research in terms of design limitations has shown that measuring the contact resistance only at selected values of the loading force provides incomplete information regarding the elastic-plastic deformation transition of a-spot. Continuously increasing the loading force and simultaneously measuring the contact resistance in synchronous mode can extremely smooth the results.

The sample holder should be designed so that one side of the sample leans against the base to eliminate its deflection. Otherwise, a part of the contact force is absorbed due to the elasticity of the material. The initially designed mounting, where the sample must be inserted through the holder, is also not optimal. The sample surface may be scratched when being inserted through the holder. Due to its dimensions, it is also difficult to ensure that the current is fed through a conductor with a sufficient cross-section so that it does not affect the sample by its heating. If pads are used along the path of the force transmission to the contact or force gauge to delimit the design tolerances, they must not be elastic (rubber) but rigid, otherwise, the applied force can be absorbed.

Numerical simulation showed that if the supply conductors are attached to the sample at one location, additional deformation of the current lines occurs inside the bar. This deformation, if the sample is short, may further affect the entry of the current lines into the contact point and increase the contact resistance. According to theoretical assumptions, the current lines should be deformed only by the entry into the constriction area of the a-spot without the influence of the current connecting method. It was found that the deformation of the current lines depends on the dimensions of the test sample, most of all on the type of material. The higher the electrical conductivity of the sample material, the faster the imposed deformation decreases by the one-sided input of the load current. FEM simulation models and analysis of the current field can illustrate this phenomenon very well.

In the first part of the work, the basic identification of the material samples was carried out by measuring the specific electrical resistance. Material types were selected which are commonly used for the construction of electrical devices and whose conductivities can be compared with catalog data or data available in the technical literature. Measurement uncertainties were also evaluated for complete identification. The final combined uncertainty with a coverage factor of $k = 2$ was then determined to obtain the resulting specific resistivity of the material. Table 6 shows that the largest uncertainty contribution (53%) is the measurement of the sample diameter D followed by the measurement of the current magnitude (37%). The remaining 10% is then accounted for by the uncertainty of the electrode voltage span and DC voltage measurements.

To reduce the uncertainty, the use of a micrometer with a high resolution, e.g., 0.1 µm, seems more appropriate. The ovality of the sample should also be checked with a higher density measurement. Similarly, the higher uncertainty in the current measurement can be attributed to the use of an inexpensive multi-meter, which was only pre-calibrated in the laboratory at a lower current value of $I = 10$ A. On the DC range, the longer measurements showed current drift caused by the internal shunt resistor of the multi-meter. This led generally to the need for long breaks during individual resistivity measurements. For further use, the resistivity measurement has already been optimized and includes the utilization of a calibrated shunt resistor and indirect current measurement using a 24-bit DAQ card.

The main focus of the work was on the measurement of contact resistance as a function of the applied force. All the materials tested showed an enormous influence of chemical layers, which increase the overall contact resistance extremely. During the measurement, the surface layers were not disturbed because the measuring system applies force only in the direction of contact without sliding movement or vibration. This was the reason why the material-dependent K_C factor for Cu was significantly higher than the tabular values used for the design of the power contacts. For power contacts, it is automatically assumed that at least some sliding motion and intensive cleaning of the surface layers occurs when the contact surfaces are coupled. Similarly, surface treatment in the form of brushing could cause an increased contact resistance value.

The influence of the surface layer was extremely pronounced in the case of aluminum bars. The material used is not the direct electrical grade of aluminum but is a structural type with a lower electrical conductivity. During the measurements, it was often the case that an infinite resistance R_{CO} was measured even with a relatively large applied force $F > 1$ N. No conductive interface formed and the surface layers were able to maintain the set potential difference $\Delta V = 1$ V. In many cases, this phenomenon occurred after the contact force had been reduced and then re-applied. A partial improvement in the measurement repeatability was achieved when the aluminum bars were resanded with fine sandpaper at the holder-sample interface and immediately fastened. The tightening of the screws in the holder had to be stronger than for the remaining samples. Overall, it was confirmed that such a holder design is completely inappropriate. In the upgraded design, this solution will be replaced by a holder with several fixing points around the circumference and a tightening belt.

The measurement of the fritting voltage was performed at a lower value of contact force, due to the maximum limits of the current source. With high contact forces or very clean Cu contact surfaces, extremely high current values of $I > 1000$ A are needed to achieve softening voltage. This is also shown in the R-V diagram of the reference by Dr. Holm and other authors, who adopted it in agreement. For real contacts and safety reasons, the voltage drop across the contacts must be many times lower than the softening voltage of the contact material. The measurements performed at a lower contact force are then an illustration of the classical fritting process of cleaning contact surfaces due to the passage of an electric current.

For alloys such as brass, the grain size of the individual material components is important. The grain size on the surface can be comparable to the size of the a-spot and can affect the contact connection. For example, for an applied force of $F = 1$ N, the a-spot diameter is approximately $2a = 90$ μm according to the parameters of Table 7. The grain size of Cu without heat treatment is in the order of 50 μm. Thus there can be an overlapping of Cu, Zn, and Pb components in the brass alloy and can cause either Zn + Zn or Cu + Cu, or Cu + Zn to be in contact. Then, the breaks in the *V-I* curve that are close to the softening voltage of Zn (0.1 V) and Cu (0.12 V) and the welding voltage of Pb (0.19 V) are quite comprehensible.

The steel samples demonstrated significant fritting when low contact forces were applied. The welding voltage was then probably reached just before the maximum load current was reached. Compared to the other tested materials, a significant force (tearing) had to be applied to separate the contact surfaces. This indicates the need for strong contact forces in potential connections with steel conductors and, due to the hardness of its surface, the use of large clamping forces in connections (typically ground rods and grounding systems).

5. Conclusions

The purpose of the article was to provide basic information with a more detailed description of the design of a test device for the validation of contact resistance of selected materials used in the design of equipment in electrical engineering.

A set of measurements of specific electrical conductivity (resistivity) of selected materials was performed and as a result, these values were refined. Values of electrical conductivity reported in the technical literature show a higher percentage of variance. The measurements of this material parameter were made concerning the uncertainty determination with an expansion factor of $k = 2$. This type of measurement procedure was created to validate newly developing materials in Al/Cu and rotary swaging combinations that may have a high potential for use in electrical engineering.

The simulation part demonstrated the enormous influence of the deformation of the current lines, neglecting the effect of the position of the input supply electrodes and the method of mounting the measured sample. It was presented that both the material of the holder and the material of the sample and its geometrical dimensions play a role in the distribution of the deformation. Based on the simulation outputs, the overall shape and dimensions of the contact holder were optimized.

The design limitations and flaws of the designed contact resistance measurement system were also found. The knowledge gained from the use of the measurement system will be further applied to the redesign of the new and improved frame structure. The newly developed system will allow additional vertical movement to simulate the sliding effect in the contacts. This could better simulate the real conditions under which the power contacts are switched. This paper shows the importance of measuring the fundamental physical properties of construction materials used in electrical engineering. Typically, the design of devices considers, among other things, the resistance of conductive paths, which in turn determines the power conversion losses and the efficiency of the system. By measuring classical or newly developed materials, we can better search for the appropriate design layout of modern electrical products.

Author Contributions: Conceptualization, P.K. and P.B.; Formal analysis, P.B. and T.M.; Investigation, P.B. and L.H.; Methodology, P.B. and L.H.; Software, T.M.; Supervision, P.K. and L.H.; Validation, P.K.; Visualization, P.K., P.B. and T.M.; Writing—original draft, P.K.; Writing—review & editing, P.K. and P.B. All authors have read and agreed to the published version of the manuscript.

Funding: This research was supported by the Ministry of Industry and Trade of the Czech Republic from the Specific Research Project no. CZ.01.1.02/0.0/0.0/20_321/0024559 and project of the Czech Science Foundation no. 22-11949S and by internal funding program SGS of the VSB—Technical University of Ostrava, identification no: SP2023/003 and no: SP2023/005.

Institutional Review Board Statement: Not applicable.

Informed Consent Statement: Not applicable.

Data Availability Statement: Not applicable.

Conflicts of Interest: The authors declare no conflict of interest. The funder had no role in the design of the study; in the collection, analyses, or interpretation of the data; in the writing of the manuscript; or in the decision to publish the results.

References

1. Brown, R.E. *Electric Power Distribution Reliability*, 2nd ed.; CRC Press: Boca Raton, FL, USA, 2008; ISBN 9780849375675.
2. Montanari, G.C. New Technologies, Grid Resiliency and Sustainable Power: It Is Also a Matter of Electrical Apparatus Reliability. In Proceedings of the 2021 IEEE Power & Energy Society Innovative Smart Grid Technologies Conference (ISGT), Washington, DC, USA, 16–18 February 2021. [CrossRef]
3. Smeets, R.; Janssen, A.; van der Sluis, L.; Peelo, D.F.; Kapetanovic, M. *Switching in Electrical Transmission and Distribution Systems*; Wiley-Blackwell: Hoboken, NJ, USA, 2014; ISBN 9781118381359.
4. Horák, A.; Prýmek, M.; Sikora, T. Power Network Reliability Computations Using Multi-Agent Simulation. In *Lecture Notes in Computer Science*; Springer: Berlin/Heidelberg, Germany, 2012; pp. 63–81. ISBN 9783642346446.
5. Erbrink, J.J.; Gulski, E.; Smit, J.J.; Seitz, P.P.; Quak, B.; Leich, R.; Malewski, R. Diagnosis of Onload Tap Changer Contact Degradation by Dynamic Resistance Measurements. *IEEE Trans. Power Deliv.* **2010**, *25*, 2121–2131. [CrossRef]
6. Faiz, J.; Siahkolah, B. *Electronic Tap-Changer for Distribution Transformers*; Springer: Berlin, Germany, 2011; ISBN 9783642199103. [CrossRef]
7. Bul, B.K.; Butkevich, G.V. *Fundamentals of the Theory of Electrical Apparatus: Textbook for Electrotechnical Faculties of Institutes of Higher Education*; HighSchool-Press: Moscow, Russia, 1970; 600p.
8. Babikov, M.A. *Electrical Apparatus. Part 1. Fundamentals of Theory*; GEI: Moscow, Russia, 1951; 420p.
9. *Electrical Contacts, Materials and Applications: Fundamentals, Technologies, Test Methods*, 2nd ed.; Vinaricky, E. (Ed.) Springer: Berlin, Germany, 2012; ISBN 9783642626982.
10. Holm, R. *Electric Contacts: Theory and Application*; Springer: Berlin, Germany, 2010; ISBN 9783642057083.
11. *Electrical Contacts: Principles and Applications*, 2nd ed.; Slade, P.G. (Ed.) CRC Press: Boca Raton, FL, USA, 2017; ISBN 9781439881316.
12. Braunovic, M.; Myshkin, N.K.; Konchits, V.V. *Electrical Contacts: Fundamentals, Applications and Technology*; CRC Press: Boca Raton, FL, USA, 2006; ISBN 9780849391088.
13. Leung, C.H.; Lee, A.; Wang, B.-J. Thermal Modeling of Electrical Contacts in Switches and Relays. In Proceedings of the Electrical Contacts—1995, the Forty-First IEEE Holm Conference on Electrical Contacts, Montreal, QC, Canada, 2–4 October 1995. [CrossRef]
14. McGowan, D. Power Connector Evaluation for Thermal Performance. In Proceedings of the 2008 Twenty-Third Annual IEEE Applied Power Electronics Conference and Exposition, Austin, TX, USA, 24–28 February 2008. [CrossRef]
15. Liu, M.; Proudhon, H. Finite Element Analysis of Contact Deformation Regimes of an Elastic-Power Plastic Hardening Sinusoidal Asperity. *Mech. Mater.* **2016**, *103*, 78–86. [CrossRef]
16. Park, S.W.; Cho, H. A Practical Study on Electrical Contact Resistance and Temperature Rise at at the Connections of the Copper Busbars in Switchgears. In Proceedings of the 2014 IEEE 60th Holm Conference on Electrical Contacts (Holm), New Orleans, LA, USA, 12–15 October 2014. [CrossRef]
17. Sawada, S.; Shimizu, K.; Hattori, Y.; Tamai, T.; Iida, K. Analysis of Contact Resistance Behavior for Electric Contacts with Plating Layer. In Proceedings of the 2010 56th IEEE Holm Conference on Electrical Contacts, Charleston, SC, USA, 4–7 October 2010. [CrossRef]
18. Li, B.; Hong, J.; Shao, G.; Du, F. An Integrated Mechanical–Electrical Predictive Model of Electrical Contact Resistance between Two Rough Surfaces. *Tribol. Trans.* **2015**, *58*, 537–548. [CrossRef]
19. Qu, J.; Li, X.; Wang, Q. Experimental Study on Electric Resistance of Tilted Contact in Air Circuit Breaker. In Proceedings of the 2016 IEEE 62nd Holm Conference on Electrical Contacts (Holm), Clearwater Beach, FL, USA, 9–12 October 2016. [CrossRef]
20. Monnier, A.; Froidurot, B.; Jarrige, C.; Meyer, R.; Test, P. A Mechanical, Electrical, Thermal Coupled-Field Simulation of a Sphere-Plane Electrical Contact. In Proceedings of the Fifty-First IEEE Holm Conference on Electrical Contacts, Chicago, IL, USA, 26–28 September 2005. [CrossRef]
21. Napieralska-Juszczak, E.; Komeza, K.; Morganti, F.; Sykulski, J.K.B.; Vega, G.; Zeroukhi, Y. Measurement of Contact Resistance for Copper and Aluminium Conductors. *Int. J. Appl. Electromagn. Mech.* **2017**, *54*, 1–13. [CrossRef]
22. Iatcheva, I.; Darzhanova, D.; Manilova, M. Modeling of Electric and Heat Processes in Spot Resistance Welding of Cross-Wire Steel Bars. *Open Phys.* **2018**, *16*, 1–8. [CrossRef]
23. Mercier, D.; Mandrillon, V.; Holtz, A.; Volpi, F.; Verdier, M.; Bréchet, Y. Quantitative Evolution of Electrical Contact Resistance between Aluminum Thin Films. In Proceedings of the 2012 IEEE 58th Holm Conference on Electrical Contacts (Holm), Portland, OR, USA, 23–26 September 2012. [CrossRef]
24. Hatakeyama, T.; Kibushi, R.; Ishizuka, M.; Tomimura, T. Fundamental Study of Surface Roughness Dependence of Thermal and Electrical Contact Resistance. In Proceedings of the 2016 15th IEEE Intersociety Conference on Thermal and Thermomechanical Phenomena in Electronic Systems (ITherm), Las Vegas, NV, USA, 31 May–3 June 2016. [CrossRef]

25. Landry, M.; Mercier, A.; Ouellet, G.; Rajotte, C.; Caron, J.; Roy, M.; Brikci, F. A New Measurement Method of the Dynamic Contact Resistance of HV Circuit Breakers. In Proceedings of the 2006 IEEE/PES Transmission & Distribution Conference and Exposition: Latin America, Caracas, Venezuela, 15–18 August 2006. [CrossRef]
26. Kulkarni, S.; Doiphode, B.; Kulkarni, Y. An Overview of Dynamic Contact Resistance Measurement of HV Circuit Breakers. In Proceedings of the 2018 IEEE/PES Transmission and Distribution Conference and Exposition (T&D), Denver, CO, USA, 16–19 April 2018. [CrossRef]
27. Runde, M.; Lillevik, O.; Larsen, V.; Skyberg, B.; Mjelve, A.; Tonstad, A. Condition Assessment of Contacts in Gas-Insulated Substations. *IEEE Trans. Power Deliv.* **2004**, *19*, 609–617. [CrossRef]
28. Stanisic, Z.; Neimanis, R. A New Ultra Lightweight Method for Static and Dynamic Resistance Measurements. In Proceedings of the 2010 IEEE International Symposium on Electrical Insulation, San Diego, CA, USA, 6–9 June 2010. [CrossRef]
29. Abdollah, M.; Razi-Kazemi, A.A. Intelligent Failure Diagnosis for Gas Circuit Breakers Based on Dynamic Resistance Measurements. *IEEE Trans. Instrum. Meas.* **2019**, *68*, 3066–3077. [CrossRef]
30. Gheorghita, C.M.; Adam, M.; Andrusca, M.; Munteanu, A.; Dragomir, A. About Contact Resistance of the Electrical Equipment. In Proceedings of the 2017 International Conference on Modern Power Systems (MPS), Cluj-Napoca, Romania, 6–9 June 2017. [CrossRef]
31. Adam, M.; Andrusca, M.; Munteanu, A.; Dragomir, A. About the Dynamic Contact Resistance of the Circuit Breakers. In Proceedings of the 2016 International Conference and Exposition on Electrical and Power Engineering (EPE), Iasi, Romania, 20–22 October 2016. [CrossRef]
32. Chen, Z.-K. Impact Wear of Electric Contact. In Proceedings of the Electrical Contacts—1991, the Thirty-Seventh IEEE HOLM Conference on Electrical Contacts, Chicago, IL, USA, 6–9 October 1991. [CrossRef]
33. Taheri, P.; Hsieh, S.; Bahrami, M. Investigating Electrical Contact Resistance Losses in Lithium-Ion Battery Assemblies for Hybrid and Electric Vehicles. *J. Power Sources* **2011**, *196*, 6525–6533. [CrossRef]
34. Jemaa, N.B.; Queffelec, J.L.; Travers, D. Apparatus and Methods for Electrical Contact Resistance Study of Cleaned and Corroded Materials. *Meas. Sci. Technol.* **1990**, *1*, 282–286. [CrossRef]
35. Queffelec, J.L.; Ben Jemaa, N.; Travers, D.; Pethieu, G. Materials and Contact Shape Studies for Automobile Connectors Development. *IEEE Trans. Compon. Hybrids Manuf. Technol.* **1991**, *14*, 90–94. [CrossRef]
36. Puyol, R.; Suarez, S. A Contact Resistance Measurement Setup for the Study of Novel Contacts. In Proceedings of the 2017 IEEE URUCON, Montevideo, Uruguay, 23–25 October 2017. [CrossRef]
37. McBride, J.W. The Relationship between Surface Wear and Contact Resistance during the Fretting of In-Vivo Electrical Contacts. *IEEE Trans. Compon. Packag. Technol.* **2008**, *31*, 592–600. [CrossRef]
38. Ren, W.; He, Y.; Jin, J.; Man, S. Investigations of the Contact Bounce Behaviors and Relative Dynamic Welding Phenomena for Electromechanical Relay. *Rev. Sci. Instrum.* **2016**, *87*, 065111. [CrossRef] [PubMed]
39. Li, L.; Deng, G.; Zhai, W.; Li, S.; Gao, X.; Wang, T. The Growth of Intermetallic Compounds and Its Effect on Bonding Properties of Cu/Al Clad Plates by CFR. *Metals* **2022**, *12*, 1995. [CrossRef]
40. Kocich, R.; Kunčická, L. Optimizing Structure and Properties of Al/Cu Laminated Conductors via Severe Shear Strain. *J. Alloys Compd.* **2023**, *953*, 170124. [CrossRef]
41. Yu, S.; Zhang, Y.; Chen, C.; Zhang, F.; Nian, H. Loss Estimation of Brushless Doubly-Fed Generator with Hybrid Rotor Considering Multiple Influence Factors. *IEEE Access* **2020**, *8*, 60043–60051. [CrossRef]
42. Hsu, J.S.; Franco Ferreira, E.A.; Coomer, C.L.; Jenkins, S.M. A New Manufacturing Technology for Induction Machine Copper Rotors. In *SAE Technical Paper Series*; SAE International: Warrendale, PA, USA, 2002.
43. Pragana, J.P.M.; Baptista, R.J.S.; Bragança, I.M.F.; Silva, C.M.A.; Alves, L.M.; Martins, P.A.F. Manufacturing Hybrid Busbars through Joining by Forming. *J. Mater. Process. Technol.* **2020**, *279*, 116574. [CrossRef]
44. Sampaio, R.F.V.; Pragana, J.P.M.; Bragança, I.M.F.; Silva, C.M.A.; Fernandes, J.C.S.; Martins, P.A.F. Influence of Corrosion on the Electrical and Mechanical Performance of Hybrid Busbars. *Int. J. Lightweight Mater. Manuf.* **2022**, *5*, 510–519. [CrossRef]
45. Wang, C.; Qu, C.; Chen, J.; Li, B.; Liu, K.; Han, Q. Performance of Aging-Treated Heat-Resistant Al–Cu–Sc Wires. *Mater. Lett.* **2022**, *315*, 132008. [CrossRef]
46. Dashti, A.; Keller, C.; Vieille, B.; Guillet, A. Novel Approach to Optimize the Mechanical Properties of Cu-Al Composite Wires. *Int. J. Mater. Form.* **2022**, *15*, 10p. [CrossRef]
47. Windred, G. Electrical Contact Resistance. *J. Franklin Inst.* **1941**, *231*, 547–585. [CrossRef]
48. Jemaa, N.B. Contacts Conduction and Switching in DC Levels. In Proceedings of the Forty-Eighth IEEE Holm Conference on Electrical Contacts, Orlando, FL, USA, 23 October 2002. [CrossRef]
49. Medved, D.; Beňa, L.; Kolcun, M.; Pavlík, M. Influence of Impurities in Electrical Contacts on Increasing the Efficiency of Energy Transmission. *Energies* **2022**, *15*, 2339. [CrossRef]
50. Kurbatov, P.A. *Fundamentals of the Theory of Electrical Apparatus*, 5th ed.; Lan-Press: St Petersburg, Russia, 2015; 592p, ISBN 9785811418008/5811418000.
51. Meaden, G.T. *IElectrical Resistance of Metals*; Springer: New York, NY, USA, 2013; ISBN 9781489957191.
52. Purkait, P.; Biswas, B.; Das, S.; Koley, C. *Electrical and Electronics Measurements and Instrumentation*; McGraw-Hill Education: Maidenhead, UK, 2013; ISBN 9781259029592.

53. Bartiromo, R.; De Vincenzi, M. *Electrical Measurements in the Laboratory Practice*, 1st ed.; Springer International Publishing: Cham, Switzerland, 2016; ISBN 9783319311005.
54. Pleština, V.; Boras, V.; Turić, H. The Measurement Uncertainty in Determining of Electrical Resistance Value by Applying Direct-Comparison Method. *Energies* **2022**, *15*, 2115. [CrossRef]
55. Shapiro-Wilk Test Calculator. Available online: www.statskingdom.comshapiro-wilk-test-calculator (accessed on 19 May 2023).
56. AALCO Copper and Copper Alloys, CW004A Sheet, Plate and Bar. Available online: https://www.aalco.co.uk/datasheets/Copper-and-Copper-Alloys-CW004A-Sheet-Plate-and-Bar_32.ashx (accessed on 19 May 2023).
57. Impol Group. Aluminium Alloy AA 2030/EN AW 2030. Available online: https://www.impol.com/app/uploads/2020/07/Aluminium-alloy-AA-2030EN-AW-2030-1.pdf (accessed on 19 May 2023).
58. ASBW Material Datasheet: CuZn39Pb3 MACHINING. Available online: https://www.asbw.pt/xms/files/02_Produtos/B14_e_B12_-_CuZn39Pb3_EN.pdf (accessed on 19 May 2023).
59. Thyssenkrupp Material Data Sheet S235Jxx/MX/TIS_11. 2016. Available online: https://ucpcdn.thyssenkrupp.com/_legacy/UCPthyssenkruppBAMXFrance/assets.files/product_pdf/carbon_flat_steel_/plates_and_slabs_carbon_steel/s235jr_1_0038_11_2016_engl.pdf (accessed on 12 June 2023).

sensors

MDPI

Article

Condition Assessment of Natural Ester–Mineral Oil Mixture Due to Transformer Retrofilling via Sensing Dielectric Properties

Hesham S. Karaman [1], Diaa-Eldin A. Mansour [2,3], Matti Lehtonen [4,*] and Mohamed M. F. Darwish [1,*]

1 Department of Electrical Engineering, Faculty of Engineering at Shoubra, Benha University, Cairo 11629, Egypt; hesham.said@feng.bu.edu.eg
2 Department of Electrical Power Engineering, Egypt-Japan University of Science and Technology (E-JUST), Alexandria 21934, Egypt; diaa.mansour@ejust.edu.eg or mansour@f-eng.tanta.edu.eg
3 Department of Electrical Power and Machines Engineering, Faculty of Engineering, Tanta University, Tanta 31511, Egypt
4 Department of Electrical Engineering and Automation, School of Electrical Engineering, Aalto University, 02150 Espoo, Finland
* Correspondence: matti.lehtonen@aalto.fi (M.L.); mohamed.darwish@feng.bu.edu.eg (M.M.F.D.)

Abstract: Mineral oil (MO) is the most popular insulating liquid that is used as an insulating and cooling medium in electrical power transformers. Indeed, for green energy and environmental protection requirements, many researchers introduced other oil types to study the various characteristics of alternative insulating oils using advanced diagnostic tools. In this regard, natural ester oil (NEO) can be considered an attractive substitute for MO. Although NEO has a high viscosity and high dielectric loss, it presents fire safety and environmental advantages over mineral oil. Therefore, the retrofilling of aged MO with fresh NEO is highly recommended for power transformers from an environmental viewpoint. In this study, two accelerated aging processes were applied to MO for 6 and 12 days to simulate MO in service for 6 and 12 years. Moreover, these aged oils were mixed with 80% and 90% fresh NEO. The dielectric strength, relative permittivity, and dissipation factor were sensed using a LCR meter and oil tester devices for all prepared samples to support the condition assessment performance of the oil mixtures. In addition, the electric field distribution was analyzed for a power transformer using the oil mixtures. Furthermore, the dynamic viscosity was measured for all insulating oil samples at different temperatures. From the obtained results, the sample obtained by mixing 90% natural ester oil with 10% mineral oil aged for 6 days is considered superior and achieves an improvement in dielectric strength and relative permittivity by approximately 43% and 48%, respectively, compared to fresh mineral oil. However, the dissipation factor was increased by approximately 20% but was at an acceptable limit. On the other hand, for the same oil sample, due to the higher molecular weight of the NEO, the viscosities of all mixtures were at a higher level than the mineral oil.

Keywords: mineral oil; natural ester oil; power transformer; dielectric properties measurement; dynamic viscosity; condition assessment; mixture oil; accelerated aging; diagnostic process

Citation: Karaman, H.S.; Mansour, D.-E.A.; Lehtonen, M.; Darwish, M.M.F. Condition Assessment of Natural Ester–Mineral Oil Mixture Due to Transformer Retrofilling via Sensing Dielectric Properties. *Sensors* **2023**, *23*, 6440. https://doi.org/10.3390/s23146440

Academic Editors: Michał Kunicki, Jan Fulneček, Pawel Rozga and Andrea Cataldo

Received: 11 May 2023
Revised: 5 July 2023
Accepted: 14 July 2023
Published: 16 July 2023

1. Introduction

Power transformers have an intrinsic role in the electrical power network. Not only at the generation stations as step-up substations but also at the consumers as step-down substations. These power transformers may be dry or oil-type transformers. In the oil-type power transformers, oil is used as an insulating and cooling medium [1,2]. The popular insulating oil used in oil-filled transformers is mineral oil (MO). Mineral oil has significant toxic effects when spilling into the soil and waterways that include the following. (1) MO has low biodegradability with harmful environmental effects. This indicates that just a small portion of oil self-degrades after being released into the environment [3]. One kilogram of oil leakage waste from a transformer renders 5 million liters of water unfit for consumption [4]. (2) MO has low flash and fire points. Once a fire hazard occurs, polycyclic aromatic hydrocarbons are released as combustion byproducts, which are toxic byproducts with an extreme threat to the environment [5]. Further, burning MO releases toxic gases into the environment [4] and produces heavy and dark smoke. (3) The cost of cleaning up oil spills is often quite high; big environmental utilities spend millions of dollars each year to do so in order to minimize their consequences. (4) MO has inadequate moisture tolerance and poor performance at high temperatures. (5) MO is a non-renewable fossil resource that might run out in the next several decades [6].

Due to the above-mentioned limitations of MO, great attention must be paid to replacing MO with an alternative oil that permits a higher degree of sustainability, is environmentally friendly, and has the same insulating and cooling properties [7]. In the last two decades, there has been an ascension in the usage of natural ester oil (NEO) as a strong alternative to mineral oil because of its high biodegradability [8], which means it easily decomposes into the soil. Hence, many researchers are oriented to study the different properties, characterizations, health index, and condition assessment of natural ester oil using various optical spectroscopy techniques [9–12]. Moreover, the NEO properties have been compared to that of MO [13,14]. On the other hand, there are some studies interested in the properties of mineral oil mixed with another insulating liquid. Perrier et al. concluded that a mixture of mineral oil with approximately 20% synthetic ester oil achieved an improvement in the dielectric properties and the aging stability without viscosity degradation compared to mineral oil alone [15]. Nadolny et al. studied the thermal properties of different mixture ratios of mineral and ester oils. According to this study, the optimal ratio of ester oil is 5% at which the heat transferability is highest and thus provides the best cooling performance for the transformer [16]. Suwarno et al. concluded that an increase in the percentage of ester contents in mineral oil provides an increase in the mixture's breakdown voltage, although the dissipation factor is slightly degraded [17]. In [18], 15% and 20% ester oil ratios provided a better dielectric behavior than other samples in this study. Toudja et al. reported that the measured charging current, resistivity, and mobility of different samples show that a mixture of 85% mineral oil with 15% plant oil is the most excellent mixture [19]. Dombek et al. presented that, with an increase in the ester contents in synthetic ester/mineral oil mixtures, the flash point and fire point determined with the open cup technique increased [20]. In [21], Beroual et al. reported that, under AC or DC applied voltage, Jatropha methyl ester oil has a higher breakdown voltage than MO, and the addition of this oil to MO extensively upgrades the breakdown voltage of the obtained mixture. Further, Ref. [22] proposes a prediction model of the transformer oil breakdown voltage in the existence of diverse barrier effects for point/plane gap systems with an AC supply voltage using a Box–Behnken design. Recently, Dixit et al. investigated the temperature distribution within the whole structure winding as well as the natural cooling category distribution transformer underneath a retrofilling with natural ester [23]. In addition, a comprehensive literature review on the mixture of MO with other alternative dielectric fluids such as natural or synthetic esters is presented [1]. This is helpful for utilities, researchers, as well as transformer owners that are interested in ester liquids besides retrofilling aspects.

As revealed in the above-mentioned literature, most of the existing studies have poor condition assessment and diagnostic tools of the dielectric properties for power transformer oil mixtures under aging processes. To solve this issue, this research work seeks to fill this literature gap by improving the condition assessment and diagnostic performance of the power transformer, specifically the properties of the oil inside that should be enhanced to produce a good insulating and cooling medium under aging. Therefore, replacing or retrofilling aged MO with fresh NEO is one of the approaches to improve power transformer performance. Due to the presence of some mineral oils absorbed by the pressboard between the transformer windings and at the bottom walls of the tank, the oil inside the tank after the replacement process is a mixture of fresh NEO with some quantities of aged mineral oil. The remaining quantities of MO range between 7% and 20% [1,24,25]. In [24], the breakdown voltage was evaluated for vegetable oil and synthetic oil mixed with MO. When using 20% MO, the drop in the AC breakdown voltage was approximately 3% and 7% for vegetable oil and synthetic oil, respectively. In [25], the AC breakdown strength of vegetable oils mixed with 10% and 30% MO was investigated. It was found that there is a slight decrement in AC breakdown strength after mixing with mineral oil compared to pure vegetable oil. This decrement attained approximately 6% and 13% for coconut oil and soybean oil, respectively, when mixed with 30% MO. Previous studies on transformer retrofilling considered only the breakdown voltage for investigation. In addition, these studies used MO in the fresh state, which does not represent the actual MO state during the retrofilling process, where MO is usually aged when retrofilling.

Therefore, in this paper, the mixture of NEO with aged MO was investigated to represent the actual condition when retrofilling. Two percentages of aged MO were considered, 10% and 20%, at different aging periods. Two different aging periods were considered, 6 days and 12 days, simulated with an accelerated aging process to be equivalent to 6 years and 12 years in the actual field service. The dielectric properties, breakdown strength, relative permittivity, and dielectric losses were sensed using a LCR meter and oil tester devices for all prepared samples to evaluate the mixtures' dielectric properties. Also, Weibull distribution analysis was introduced to evaluate the probability of breakdown for the different mixtures. The viscosity of all prepared samples was measured to investigate the dynamic performance of the different mixtures. In addition, the physical mechanism is discussed to present the dielectric and dynamic performance of ester oil mixed with mineral oil in a power transformer. Finally, an electrostatic model of an oil-filled distribution transformer is developed with COMSOL Multiphysics Software to illustrate the electric field distribution inside the transformer in the case of refilling the transformer with natural ester oil (for the best retrofilling oil sample).

2. Oil Specifications and Experimental Proceedings

In this section, the specifications of NEO and MO are presented. Then, the MO aging process and the preparation of samples are described. Finally, the sensing system used for evaluating the dielectric properties in the laboratory is detailed.

2.1. Oil Specifications

The obtained samples in this study were prepared by mixing fresh natural ester oil and mineral oil with different percentages, noting that the NEO and MO are commercial types and treated by the manufacturer. Moreover, all samples were treated with a vacuum oven in the laboratory to remove the moisture before starting the experimental process. Table 1 presents the specifications of both oils based on the datasheets provided by the suppliers.

Table 1. Specifications of natural ester oil and mineral oil.

Characteristic	Specification	
Oil type	MO	NEO
Appearance	Clear, transparent	Clear, light green
Density at 20 °C	0.88 (kg/cm^3)	0.92 (kg/cm^3)
Flash point	140 (°C)	325 (°C)
Pour point	−57 (°C)	−21 (°C)
Electric strength	>60 (kV)	>60 (kV)
Viscosity at 40 °C	9.4 (mm^2/s)	<40 (mm^2/s)
Total acidity	<0.01 (mg KOH/g)	<0.06 (mg KOH/g)
Water content	<20 (ppm)	<200 (ppm)

2.2. Aging Process

In order to prepare aged mineral oil that simulates the aged oil in the power transformer in service, an accelerated thermal aging process is required [26]. This process was carried out for the fresh mineral oil using a heating oven. At 120 °C, one day in the oven is equivalent to 380 days, i.e., approximately one year in the field [27], where the base operational transformer oil temperature is 60 °C and the aging rate of oil is doubled when the temperature raises by 7 °C, according to the following Equations (1) and (2) [11]:

$$\textbf{\textit{Aging accelerating factor}} = \frac{\mathbf{120\ ^{\circ}C - 60\ ^{\circ}C}}{\mathbf{7\ ^{\circ}C}} = \mathbf{8.57} \quad (1)$$

$$\textbf{\textit{Time factor}} = \mathbf{2^{8.57}} = \mathbf{380} \quad (2)$$

In this study, two different aging periods were considered. These periods were 6 days and 12 days that were equivalent to 6 years and 12 years in the field. The obtained aged mineral oil samples aged for 6 days and 12 days were called MO6D and MO12D, respectively.

2.3. Sample Preparation

The mineral oil was aged as mentioned in the previous subsection to simulate the real field. The aged mineral oil was mixed with fresh natural ester oil after heating it in a vacuum oven at 80 °C for 48 h to remove gas bubbles and moisture contents. A magnetic stirrer was used to mix the two types of oils at 600 rpm for 30 min. The prepared oil samples were left for 24 h before the measuring processes as shown in Figure 1.

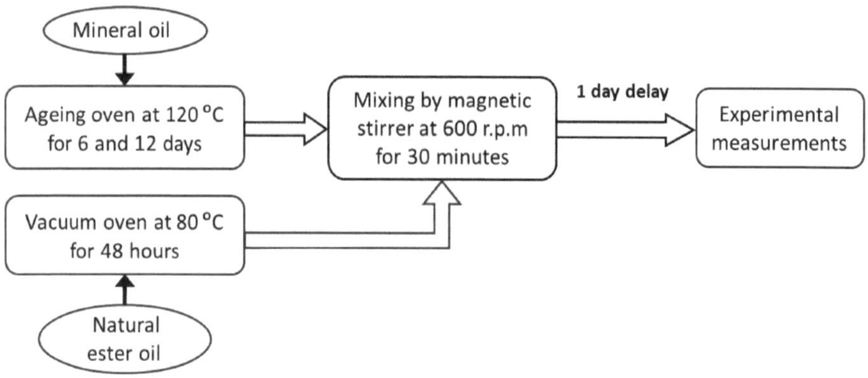

Figure 1. Schematic diagram to illustrate the preparation process of the oil samples.

2.4. Dielectric Strength

The dielectric strength of the transformer oil is the most important characteristic that indicates the quality of its insulation property. Consequently, when the dielectric strength is high, it is an indication of a good quality transformer oil that reflects positively on the transformer's performance. Therefore, in this study, the prepared oil samples were subjected to a breakdown test using the AC oil tester presented in Figure 2 (from 0 to 100 kV) to evaluate and sense the dielectric strength of each sample. The test proceeded based on the IEC-60156 standard with a 2 kV/s voltage ramp rate and a 2.5 mm separation distance between the mushroom-shape test electrodes. For each sample, ten results were recorded and analyzed with special sensors to perform the Weibull distribution cumulative probability function to provide all probabilities of AC breakdown strength with a small number of tests [28]. The Weibull distribution cumulative probability function for the oil sample $F(v)$ can be introduced as shown in Equation (3): v is the breakdown voltage in kV, λ denotes the scale parameter in kV, and ξ represents the shape parameter [29].

$$F(v) = 1 - e^{-(\frac{v}{\lambda})^{\xi}} \tag{3}$$

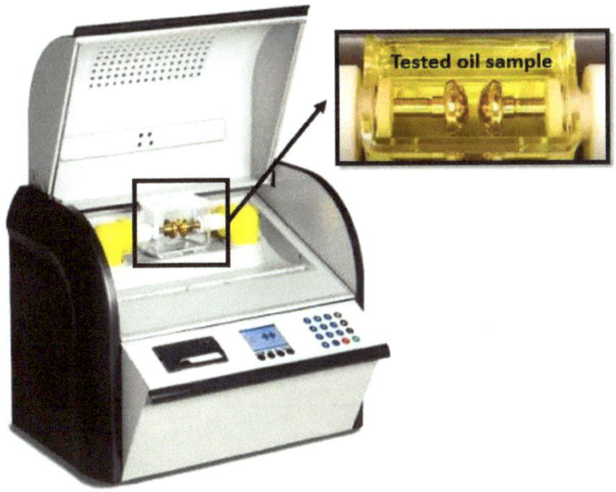

Figure 2. AC breakdown tester for liquid insulations.

2.5. Dielectric Properties

Due to the presence of an electric field on the insulating oil, dielectric polarization occurs for its molecules. The ability of polarization under the effect of the applied field can be defined with the expression of the relative permittivity (ε_r) [30]. In the composite insulation system of the oil-filled transformer (oil/paper), the greater the relative permittivity of the insulating oil, the more uniform the electric field at the interface between the oil and the insulating paper. Therefore, increasing the relative permittivity of the insulating oil is preferable for good performance of the transformer [29]. On the other hand, there is another parameter that indicates the insulation system quality. This parameter is the dissipation factor of the insulating liquid (tan δ). A high value of the dissipation factor refers to the presence of more existing contaminations and imperfections in the insulating oil [31,32]. Therefore, mitigation of the dissipation factor value provides a positive impact on the insulation system. The relative permittivity and dissipation factor can be estimated from the following Equations (4)–(7); Table 2 defines the different variables in the given equations.

$$\varepsilon_r = \varepsilon' - j\varepsilon'' \tag{4}$$

$$\tan\delta = \frac{\varepsilon''}{\varepsilon'} \tag{5}$$

$$\varepsilon' = \frac{dC_p}{\varepsilon_0 A} \tag{6}$$

$$\varepsilon'' = \frac{d}{2\pi f \varepsilon_0 A R_p} \tag{7}$$

In order to sense and evaluate the relative permittivity values as well as the dissipation factor of the oil sample, the capacitance (*Cp*) and the resistance (*Rp*) of the oil sample must be measured. The sensing system refers to the experimental measurements in the laboratory used for evaluating the dielectric properties. In this study, the Agilent E4980A Precision LCR meter was employed to sense and measure the *Cp* and *Rp* values using advanced sensors built into the LCR meter under the adjustment of the equivalent circuit parallel model of the LCR meter. Figure 3 presents the LCR meter and the test cell that were used in this test with 0.3 mm gap spacing and 38 mm electrode diameter under a wide range of frequencies from 20 Hz to 1 MHz. After measuring the values of Cp and Rp of each sample with the LCR meter, the value of Cp was applied in Equation (6) to evaluate the real permittivity ε', and the value of Rp was employed in Equation (7) to estimate the imaginary permittivity ε''. Thus, the obtained values of relative permittivity and the dissipation factor were computed from Equations (4) and (5), respectively.

Table 2. Definitions of the variables of the relative permittivity (ε_r) and dissipation factor (tan δ) equations.

Variable	Definition
ε'	Dielectric coefficient
ε''	Dielectric loss
d	Gap spacing between the test electrodes
C_p	Oil sample capacitance
ε_o	Permittivity of free space
A	Test electrode cross-sectional area
f	Frequency
R_p	Oil sample resistance

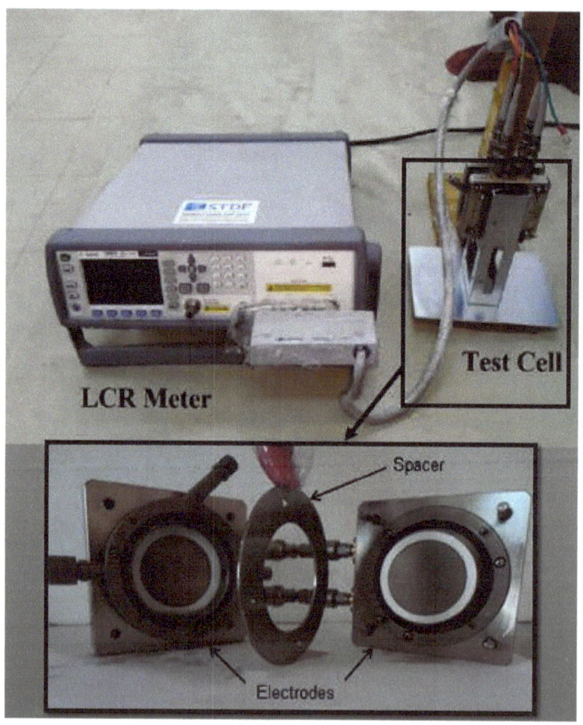

Figure 3. LCR meter and the test cell used for the dielectric measurements.

2.6. *Dynamic Viscosity Preparation*

It is known that the viscosity of any liquid is the resistance of this liquid to flow. The viscosity of the transformer oil is a very important property in oil-filled transformers for providing a good heat-exchange process. Therefore, a lower viscosity of the oil results in a higher circulation speed of the oil and better efficiency of the cooling system. In this study, the Kinematic Viscosity Bath (KV3000) was used to measure the viscosity of the prepared oil samples. This device is used to sense and measure the flow of the liquid under gravity or vacuum at specified controlled temperatures.

3. Results and Discussion

3.1. Dielectric Strength

The usage of natural ester oil in power transformers not only provides environmental advantages over mineral oil but also provides higher dielectric properties. Figure 4 displays the cumulative probability function for fresh MO and fresh NEO. As shown in Figure 4, the dielectric strength of NEO is higher than that of MO by more than 50%. To study the refilling of the transformer with fresh natural ester oil, six cases of oil testing measurements were introduced in addition to the cases of fresh oils (MO and NEO). These cases are presented in Table 3, where 90% and 80% of NEO mixed with 10% and 20% of aged MO, respectively, at different aging periods are utilized. These percentages are determined based on the expected remaining quantities of MO during the transformer retrofilling process, as specified in previous studies [1,24,25].

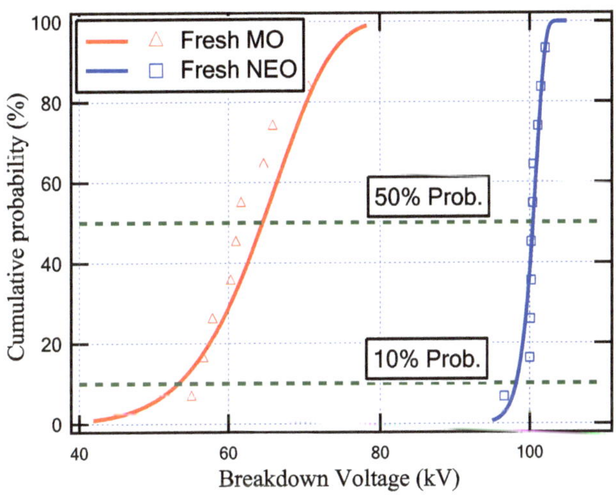

Figure 4. Cumulative probability against breakdown voltage for fresh mineral oil (MO) and fresh natural ester oil (NEO).

Table 3. Description of the oil-tested samples.

ID	Description
Case (1)	Pure mineral oil aged for 6 days (MO6D)
Case (2)	Pure mineral oil aged for 12 days (MO12D)
Case (3)	Mixture of MO aged for 6 days 10% + fresh NEO 90% (10% MO6D + 90% NEO)
Case (4)	Mixture of MO aged for 6 days 20% + fresh NEO 80% (20% MO6D + 80% NEO)
Case (5)	Mixture of MO aged for 12 days 10% + fresh NEO 90% (10% MO12D + 90% NEO)
Case (6)	Mixture of MO aged for 12 days 20% + fresh NEO 80% (20% MO12D + 80% NEO)

Figure 5 indicates the cumulative probability function for the oil samples illustrated in Table 3. In addition, Table 4 depicts the shape parameter (ξ), scale parameter (λ), breakdown voltage (BDV) with an AC supply at 50% probability (BDV50%), and BDV at 10% probability (BDV10%) for all insulating oil samples.

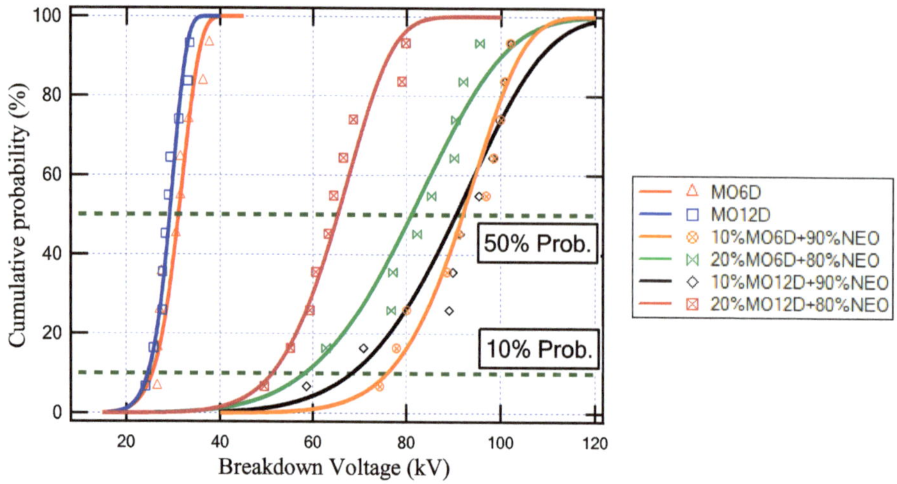

Figure 5. Cumulative probability against breakdown voltage for several aged oil samples.

Table 4. Weibull distribution analysis results for all insulating oil samples.

Oil Sample	ξ	λ (kV)	BDV 50% (kV)	BDV 10% (kV)
Fresh MO	9.8	67	64.6	53.3
Fresh NEO	83.6	101	100.4	98.2
MO6D	9.3	32.4	31.2	25.5
MO12D	10.8	30.2	29.2	24.6
10% MO6D + 90% NEO	**9.8**	**95.5**	**92**	**76**
20% MO6D + 80% NEO	5.7	86.5	81.1	58.2
10% MO12D + 90% NEO	6.7	95.7	90.5	68.3
20% MO12D + 80% NEO	7.7	68.5	65.3	51.2

The aging of any insulating oil causes a degradation of its dielectric properties due to the thermal stresses affecting the insulating oil bonds. From the obtained results, the aging of mineral oil for 6 days and for 12 days caused a decrease in the breakdown voltage at 50% probability from 64.6 kV to 31.2 kV and 29.2 kV, respectively, while the BDV at 10% probability decreased from 53.3 kV to 25.5 kV and 24.6 kV, respectively. On the other hand, the obtained results show that, due to the increase in the ratio of NEO that fills the transformer, the breakdown strength of the insulating oil inside the transformer increased as presented in Table 4. Furthermore, the oil sample that has a higher BDV is (10% MO6D + 90% NEO) as highlighted in Table 4. This sample achieved an improvement in the BDV by 43% compared to MO and by 195% compared to MO6D, which was attributed to this sample having a larger percentage of NEO and a lower aging period.

3.2. Dielectric Properties

The Agilent E4980A Precision LCR meter sensed and measured the capacitance and resistance of the different oil samples to evaluate their dielectric parameters at a frequency range from 20 Hz to 1 MHz at room temperature (25 °C). The changing of the dielectric permittivity refers to the ionic and dipolar polarizations that occur via this range of frequencies [33]. Figure 6 shows the relative permittivity versus frequency for fresh MO and fresh NEO. As presented, the relative permittivity of NEO is higher than that of MO for all frequency ranges. Regarding the dissipation factor that is shown in Figure 7, the dissipation factor of MO is lower than that of NEO by approximately 0.074. On the other hand, with regard to the longer period of aging for insulating MO, the relative permittivity was enhanced, but the dissipation factor degraded. This result is similar to that obtained with the mixture of NEO and MO as summarized in Table 5. Moreover, Figures 8 and 9 illustrate the variation in relative permittivity and the dissipation factor with different frequencies, respectively. The oil sample that has a higher BDV is (10% MO6D + 90% NEO) as highlighted in Table 4 and mentioned in the previous subsection. Furthermore, the sample of (10% MO6D + 90% NEO) achieved the best relative permittivity with a value equal to 2.829 compared to all other oil samples. Even though the dissipation factor of this oil sample is not the lowest, it is still under the acceptable value mentioned in IEC 60296.

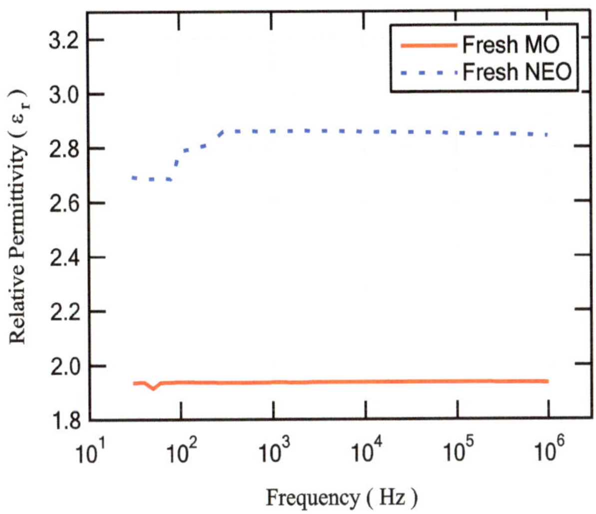

Figure 6. Relative permittivity versus frequency for fresh mineral oil (MO) and fresh natural ester oil (NEO).

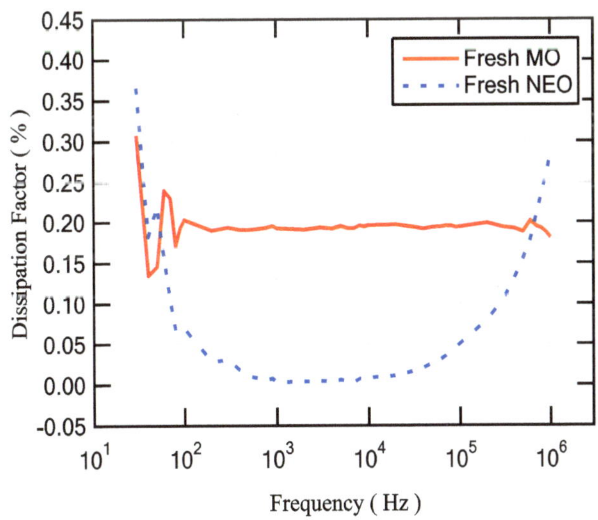

Figure 7. Dissipation factor versus frequency for fresh mineral oil (MO) and fresh natural ester oil (NEO).

Table 5. Dielectric properties for all insulating oil samples at a frequency of 50 Hz.

Oil Sample	ε_r	tan δ (%)
Fresh MO	1.914	0.146
Fresh NEO	2.685	0.220
MO6D	1.925	0.149
MO12D	2.026	0.166
10% MO6D + 90% NEO	2.829	0.179
20% MO6D + 80% NEO	2.720	0.197
10% MO12D + 90% NEO	2.814	0.169
20% MO12D + 80% NEO	2.716	0.191

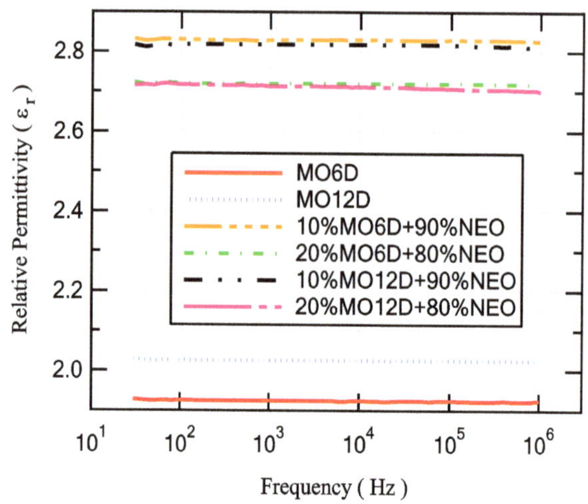

Figure 8. Relative permittivity versus frequency for several aged oil samples.

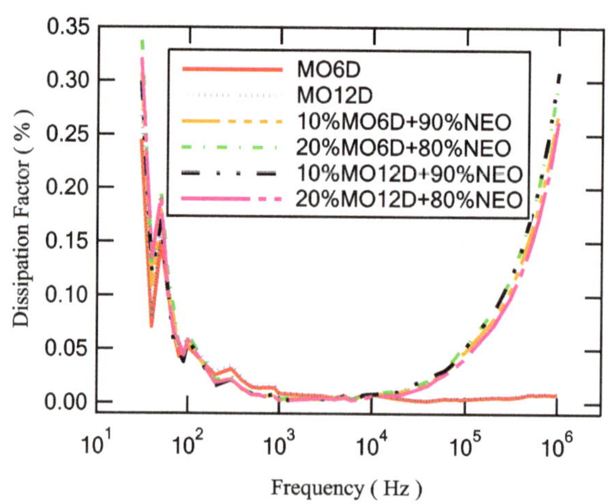

Figure 9. Dissipation factor versus frequency for several aged oil samples.

3.3. *Dynamic Viscosity Measurement*

For the enhancement of the cooling property of the insulating oil due to its higher circulation speed, a lower viscosity of the oil is necessary. Figure 10 shows the relationship between viscosity and temperature for fresh NEO and MO. The test was carried out at temperatures of 25, 50, and 75 °C. It is clear that the viscosity of MO is lower than the viscosity of NEO for all tested temperatures. This indicates that the cooling performance of MO is better than that of NEO. This is because the main component of NEO is a triglyceride, and its molecular weight is higher than that of the hydrocarbon chains of MO, so the NEO kinematic viscosity is greater than that of MO [34].

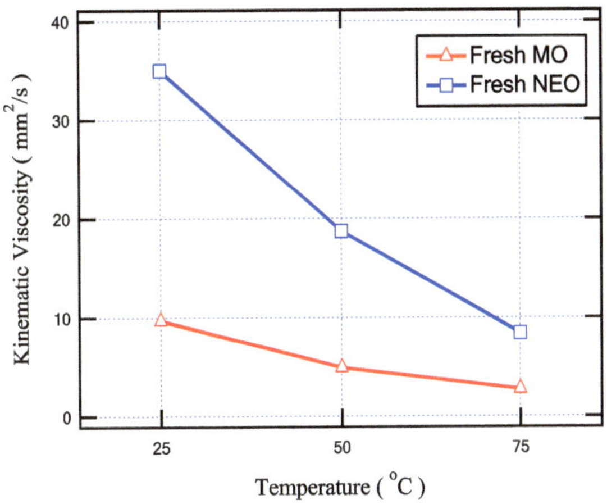

Figure 10. The variation in fresh insulating oil viscosity with temperature increment.

Due to the aging of the mineral oil for 6 and 12 days, as introduced in Figure 11, the oil viscosity increased by approximately 4% and 40%, respectively, compared with fresh MO at 50 °C. On the other hand, due to the transformer retrofilling process with NEO, the oil mixture sample that achieved the lowest viscosity is the sample of (20% MO6D + 80% NEO). This sample provides better cooling performance due to a lower aging period in addition to a lower ratio of NEO. However, the sample of (10% MO6D + 90% NEO), which achieves better dielectric performance as discussed in the previous sections, provides a good viscosity value that is close to the viscosity value of the sample of (20% MO6D + 80% NEO).

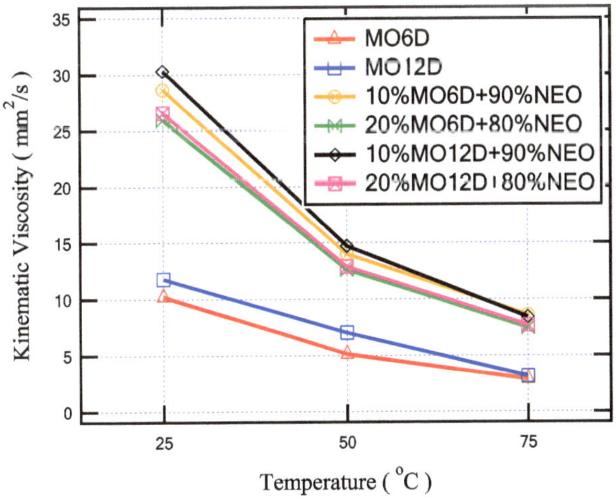

Figure 11. The variation in viscosity with temperature increment for all insulating aged oil samples.

4. Physical Mechanisms

Regarding the dielectric strength of the insulating oils, the AC-BDV of insulating oil is mainly related to the relative moisture contents. Therefore, the higher saturation moisture content of mixed insulation oil is a major aspect that produces a higher AC-BDV than that of MO at a similar absolute moisture content [35,36]. This concept can be an explanation for the increasing AC-BDV as the percentage of NEO increases in the oil mixture and also with a lower aging period as summarized in Table 4. The higher aging period of MO produces lower dielectric properties for this oil. This can be attributed to the breaking of some MO molecules due to the applied electric field during the transformer's continuous operation. From the obtained results, for 20% MO, the aging of MO for 12 days achieved an AC-BDV of 65.3 kV, which is lower than 81.1 kV achieved at 6 days, while for 10% MO, the AC-BDV was reduced from 92 kV to 90.2 kV from 6 days to 12 days aging, respectively. Note that the supply of the measured breakdown voltage here is an AC supply; otherwise, from the literature, the NEO performance under impulse waveforms or other fast transients tends to be lower than that of MO, especially if used in barrier-style insulation or homogenous paper oil insulation. Therefore, it can be interesting to study this performance in future work.

On the other hand, the relative permittivity and the dissipation factor of the mixed insulating oil are not only dependent on the percentage of NEO but also on the aging period as presented in Table 5. This dependency relates to the dipole polarization inside the mixed oil [29]. Moreover, the ionic compounds in the aged oil may be dissolved in NEO because of their polar nature; in addition, when the percentage of NEO is increased, this effect becomes prevalent. The increasing dissipation factor of the mixed oil may be due to the aged MO containing a high quantity of polar colloidal components and odorous hydrocarbons [37].

Regarding the dynamic viscosity of the insulating oils, the lower viscosity of MO for all ranges of temperature refers to a better coolant medium. This behavior may be attributed to its lighter hydrocarbon molecular weight than that inside the mixed oil that contains a large amount of NEO's heavy triglycerides [34].

5. Electrostatic Modelling of Oil-Filled Transformer

In this section, an electrostatic model of a 1 MVA, 50 Hz, 22/0.4 kV step-down transformer is simulated using COMSOL Multiphysics Software. The aim of this simulation is to evaluate the distribution of the electrostatic field inside the transformer when using the NEO–MO oil mixtures after the retrofilling process. The case of the oil mixture (10% MO6D + 90% NEO) was considered for the electrostatic field analysis, as it exhibited superior dielectric properties.

COMSOL software is utilized to resolve the desired non uniform field via the numerical finite element method to effectively model the electrostatic field inside any part of the transformer, specifically the sharp edges and interface points. The input parameters for this model involve the measured r.m.s rates of the AC breakdown voltage followed by the measured dielectric loss and permittivity of the prepared oil samples (NEO–MO mixture). In addition, the other detailed model parameters of the transformer mentioned in Table 6 and the datasheet parameters of numerous oil types are extracted from real experimental data. The computation of the two-dimensional (2D), axis-symmetric field distribution can be exploited via the finite element method (FEM) through the interface area between the winding and NEO–MO mixture. The mechanism of the COMSOL operation depends on the field region that should first be divided into smaller triangle elements to shrink the energy throughout the wide field area of interest by applying the numerical FEM [38]. Once a stationary electric field is utilized for a liquid dielectric substance, the Equations (8) and (9) deliver the electrical energy stored inside the full volume of the area under study, taking into consideration the cylindrical model geometry as well as the Laplacian field Equation (8) for static field [39].

$$\nabla^2 V = 0 \tag{8}$$

$$\vec{E} = -\nabla V \tag{9}$$

where these equations should be utilized at the unknown potential nodes to compute the electrostatic potential V; after that, the electric field strength E can be easily calculated with Equation (9).

Table 6. Technical data for the modeled transformer [40].

Item	Specs
Type	Oil-filled
Standard specification	IEC 60076
Rated output at 45 °C ambient (kVA)	1000
Method of cooling	ONAN
Oil type	Mineral oil
System of connection	DYn-11
Turns ratio	22,000/400
Resistance/phase for primary RHV (Ω)	4.23
Resistance/phase for secondary RLV (Ω)	5.11×10^{-4}
Permissible symmetrical S.C current at L.V side terminals for 2 s (kA)	62.55
Primary voltage at normal tapping (V)	22,000
Corresponding secondary voltage at no load (V)	400
Full load current at LV side I_{LV} (A)	1443.4
Full load current at HV side I_{HV} (A)	26.24
Efficiency at (100% −75%) rated output (%)	98.9–99.1
Iron losses (W)	1222
Copper losses (W)	9450
Temperature rise at rated output above 45 °C ambient temperature	
At oil top level (°C)	45
Winding temperature (°C)	55
Core temperature (°C)	55

Table 6 presents the technical data for the modeled transformer [40]. Based on that, the building of an accurate model of the transformer is very difficult, so some simplifications were considered in this work: Only one winding was considered, and the tank in the model was considered to be a perfect cylinder equivalent to one-third of the real transformer tank [41]. Moreover, an approximately 10% voltage drop across transformer windings was considered due to the copper resistance of these windings that can slightly change the electric field stress along the winding [10]. Hence, from Sections 2 and 3, all measured parameters of the NEO–MO mixture are used as input data for the simulation model to evaluate the distribution of the electrostatic field inside the transformer when using these oil mixtures.

When COMSOL Multiphysics is started, the model navigator enables the user to begin the modeling process and control all program settings. The space dimension is selected as 2D Axis-symmetric, and the application mode of electrostatic is adopted to begin working on the considered model. The model was created based on the dimensions shown in Figure 12. Moreover, Figure 13 presents the model after creation with the space dimension selected as 2D Axis-symmetric. Due to the build of the model under the electrostatic study [42], the AC voltage distribution inside all parts of the transformer can be obtained as shown in Figure 14. Furthermore, the electric field distribution can be evaluated based on the FEM using finer grid meshes as presented in Figure 15. Additionally, Figure 16 introduces the electric field distribution inside the modeled transformer, specifically at lines AA′ and BB′ and at points a, b, and c (see Figure 13). From the obtained results, it is clear that the electric field's highest value points are concentrated between the low-voltage (LV) and high-voltage (HV) windings. These stresses are due to the higher voltage of HV winding compared to LV winding.

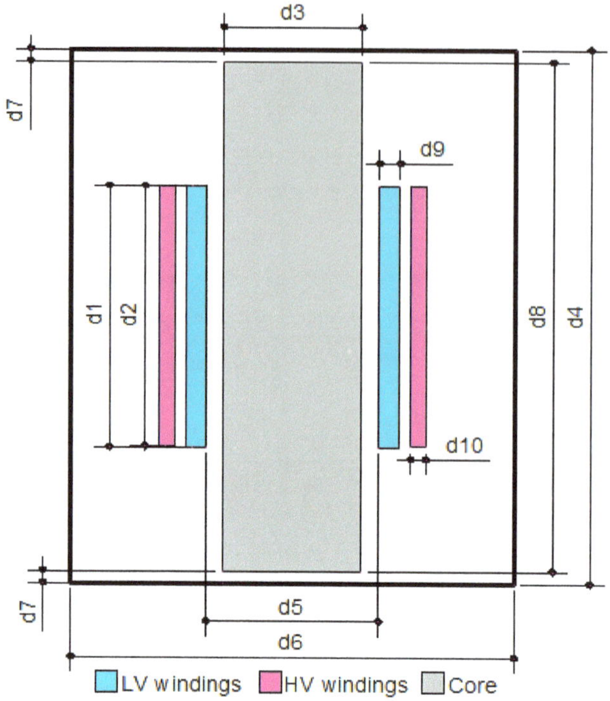

d	mm
d1	595
d2	593
d3	304
d4	1215
d5	380
d6	968
d7	28
d8	1159
d9	45
d10	33

Figure 12. Two-dimensional geometry of the studied model (all dimensions in mm).

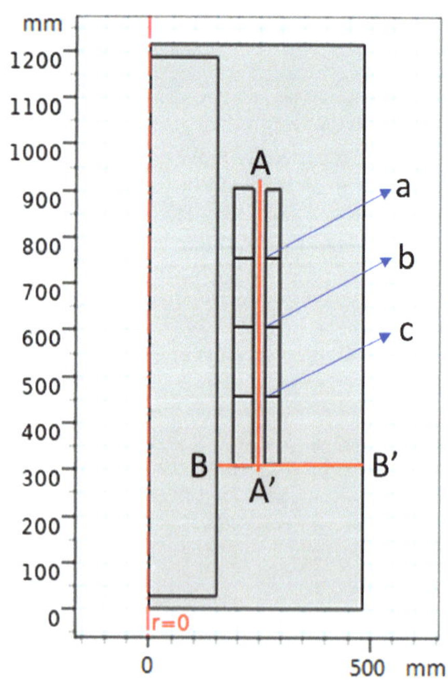

Figure 13. View of geometry creation of the model on COMSOL Multiphysics with millimeter dimensions for points (a, b, c) and line sections (AA′ and BB′).

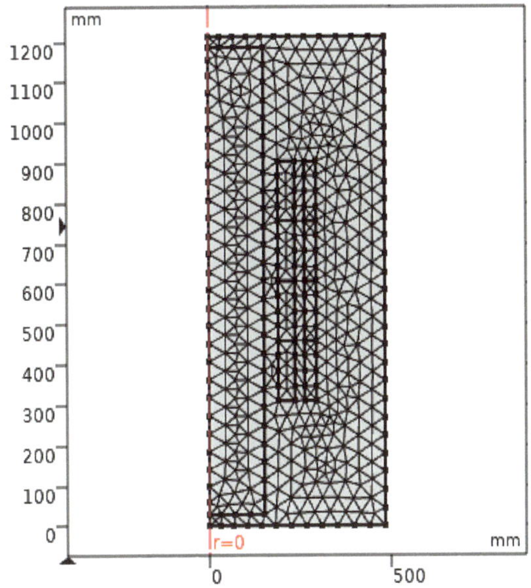

Figure 14. Model after the creation of the meshes for the finite element method solution (finer meshes style).

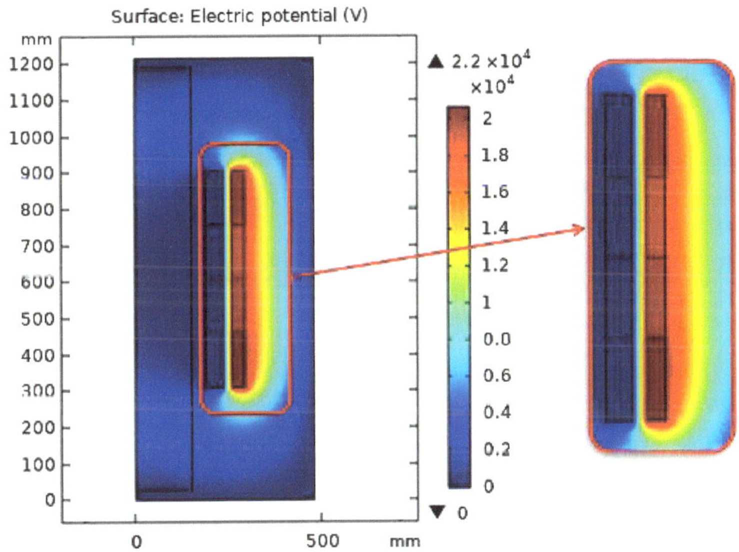

Figure 15. AC potential distribution inside all transformer parts based on FEM.

To illustrate the values of the electric field in many locations on the insulating oil, the distribution of the electric field is evaluated along the lines AA′ and BB′ presented in Figure 13. The obtained electric field values along these lines are presented in Figure 17. Based on the obtained results, along AA′ that lies midway between the HV and LV windings, the higher electric field is presented nearly to the edges of the transformer windings at vertical arc lengths of 310 mm and 905 mm with electric fields of 705 kV/m and 678 kV/m, respectively. On the other hand, along BB′, the higher electric field is presented at a horizontal arc length of 251 mm with an electric field of 706 kV/m. Hence, the highest value points of the electric field on the insulating oil are presented at the edges of the transformer windings due to the direct interface between the transformer winding and the insulating oil molecules in this region. Accordingly, the best insulating oil is (10% MO6D + 90% NEO) from the transformer retrofilling point of view, which produces a good insulation and cooling performance via these highest value points compared to the other oil samples. Moreover, the transformer will have a longer lifetime, better performance, and fewer technical problems. Moreover, Figure 18 introduces the electric field at points a, b, and c (see Figure 13) that present a slight variation in the electric field due to the voltage drop along the transformer windings. The obtained results concluded that the electric field was varied at the selected points a, b, and c with values of 615, 636, and 660 kV/m, respectively, which do not exceed 10% variations in the electric field stress along the transformer winding.

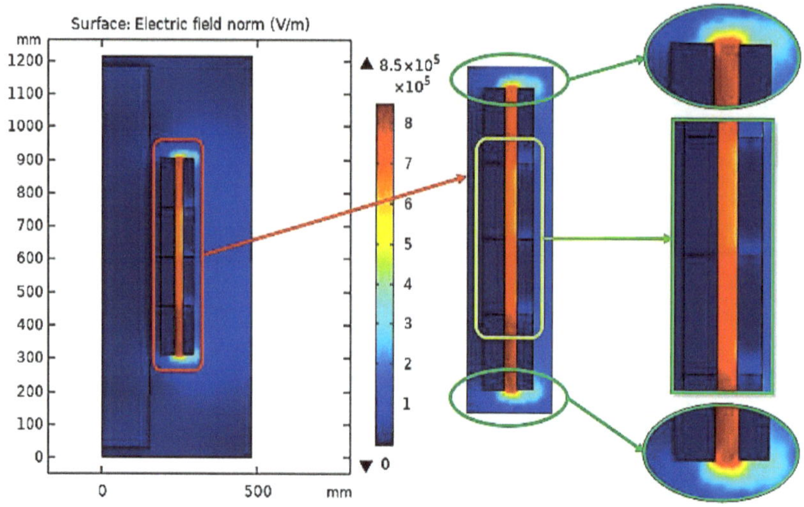

Figure 16. Electric field distribution inside the modeled transformer based on FEM.

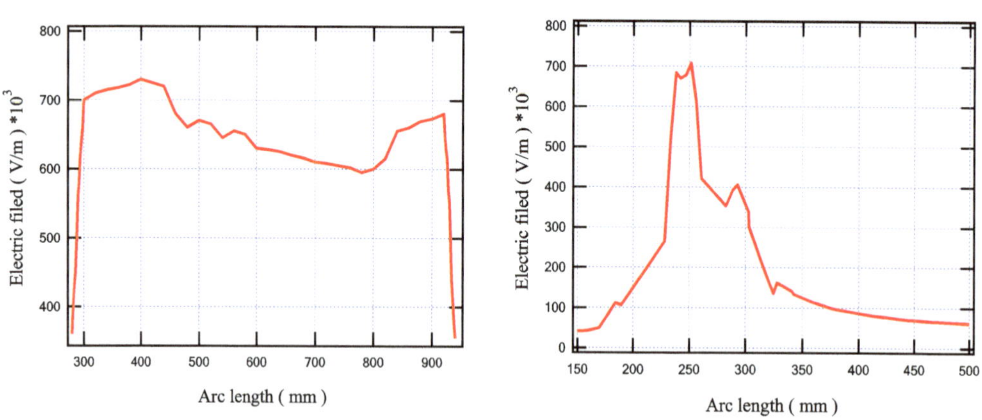

(a) (b)

Figure 17. Electric field inside the modeled transformer along (**a**) line AA′ and (**b**) line BB′.

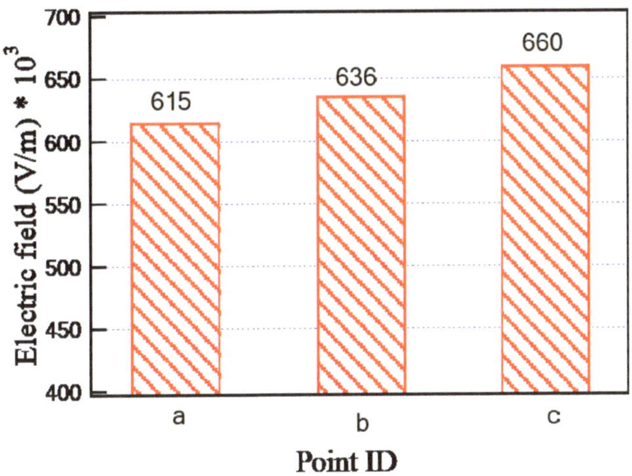

Figure 18. Electric field inside the modeled transformer at points a, b, and c.

6. Conclusions

This paper studied the dielectric properties and coolant behavior based on the dynamic viscosity of the transformer oil due to the replacement of worn-out mineral oil with natural ester oil. The experimental setup, measurements, and electrostatic simulation were introduced. From the above-mentioned results, the following remarks can be concluded:

- The mixing of NEO with aged MO leads to an enhancement in some characteristics of the mixed insulating oil, such as the dielectric strength and relative permittivity.
- The aged mineral oil has degraded dielectric and coolant properties when compared with the fresh mineral oil because of the electrical as well as thermal stresses on the oil molecules and bonds.
- The larger the amount of NEO in the mixed insulating oil, the higher the BDV due to the higher dielectric strength of the NEO that was mixed with the aged MO.
- The values of the dissipation factor for the mixed insulating oils exhibited a slight increase, but their values were kept under acceptable limits.
- The relative permittivity values for the mixed insulating oils were increased due to the increase in the percentage of NEO, which was due to the high dipole polarization inside the mixed oil.
- The dynamic viscosity of MO is lower than that of NEO for all ranges of temperature. This indicates that the cooling performance of MO is better than that of NEO.
- The electrostatic simulation of the oil transformer introduced the electric field on the insulating oil with higher values at the edges of the transformer windings due to the direct interface between the transformer winding and the insulating oil molecules in this region. Also, the variation in electric field stress on the transformer winding was taken into consideration for a more accurate model that did not exceed 10%.
- The best insulating oil (10% MO6D + 90% NEO) produces an optimal parameter for insulation and cooling performance compared to the other oil samples. Moreover, the transformer would have a longer lifetime, better diagnostic performance, and fewer technical problems.

In future work, the oil mixture performance under impulse waveforms or other fast transients will be investigated; in addition, the breakdown analysis at the paper/pressboard oil interface will be studied.

Author Contributions: All authors have contributed to the preparation of this manuscript. Conceptualization, D.-E.A.M., H.S.K., M.L. and M.M.F.D.; Methodology, D.-E.A.M., H.S.K. and M.M.F.D.; Software and Validation, D.-E.A.M., H.S.K. and M.M.F.D.; Formal Analysis and Investigation, D.-E.A.M., H.S.K. and M.M.F.D.; Resources and Data curation, D.-E.A.M., H.S.K. and M.M.F.D.; Writing—original draft preparation, D.-E.A.M., H.S.K. and M.M.F.D.; Writing—review and editing, D.-E.A.M., H.S.K., M.L. and M.M.F.D.; Visualization, D.-E.A.M., H.S.K., M.L. and M.M.F.D.; Supervision, D.-E.A.M. and M.L. All authors have read and agreed to the published version of the manuscript.

Funding: This paper is created upon work supported by Science, Technology & Innovation Funding Authority (STIFA), Egypt, under a grant (31292).

Institutional Review Board Statement: Not applicable.

Informed Consent Statement: Not applicable.

Data Availability Statement: The data presented in this study are available on request from the corresponding author.

Conflicts of Interest: The authors declare no conflict of interest.

References

1. Lyutikova, M.; Korobeynikov, S.; Rao, U.M.; Fofana, I. Mixed Insulating Liquids with Mineral Oil for High-Voltage Transformer Applications: A Review. *IEEE Trans. Dielectr. Electr. Insul.* **2022**, *29*, 454–461. [CrossRef]
2. Prasojo, R.A.; Putra, M.A.A.; Ekojono; Apriyani, M.E.; Rahmanto, A.N.; Ghoneim, S.S.; Mahmoud, K.; Lehtonen, M.; Darwish, M.M. Precise transformer fault diagnosis via random forest model enhanced by synthetic minority over-sampling technique. *Electr. Power Syst. Res.* **2023**, *220*, 109361. [CrossRef]
3. Jacob, J.; Preetha, P.; Krishnan, S.T. Review on natural ester and nanofluids as an environmental friendly alternative to transformer mineral oil. *IET Nanodielectr.* **2020**, *3*, 33–43. [CrossRef]
4. Rozga, P. Properties of new environmentally friendly biodegradable insulating fluids for power transformers. *Eur. Sci. J.* **2013**, *9*, 358–364.
5. Li, C.; Zhang, X.; Wang, X.; Zhang, X.; Liu, S.; Yuan, T.; Qu, W.; Zhang, Y. Distribution Characteristics and Potential Risks of Polycyclic Aromatic Hydrocarbon (PAH) Pollution at a Typical Industrial Legacy Site in Tianjin, North China. *Land* **2022**, *11*, 1806. [CrossRef]
6. Srivastava, M.; Goyal, S.K.; Saraswat, A. Ester oil as an alternative to mineral transformer insulating liquid. *Mater. Today Proc.* **2021**, *43*, 2850–2854. [CrossRef]
7. Ward, S.A.; El-Faraskoury, A.; Badawi, M.; Ibrahim, S.A.; Mahmoud, K.; Lehtonen, M.; Darwish, M.M.F. Towards Precise Interpretation of Oil Transformers via Novel Combined Techniques Based on DGA and Partial Discharge Sensors. *Sensors* **2021**, *21*, 2223. [CrossRef] [PubMed]
8. Mehta, D.M.; Kundu, P.; Chowdhury, A.; Lakhiani, V.K.; Jhala, A.S. A Review on Critical Evaluation of Natural Ester vis-à-vis Mineral Oil Insulating Liquid for Use in Transformers: Part 1. *IEEE Trans. Dielectr. Electr. Insul.* **2016**, *23*, 873–880. [CrossRef]
9. Badawi, M.; Ibrahim, S.A.; Mansour, D.-E.A.; El-Faraskoury, A.A.; Ward, S.A.; Mahmoud, K.; Lehtonen, M.; Darwish, M.M.F. Reliable Estimation for Health Index of Transformer Oil Based on Novel Combined Predictive Maintenance Techniques. *IEEE Access* **2022**, *10*, 25954–25972. [CrossRef]
10. Darwish, M.M.F.; Hassan, M.H.A.; Abdel-Gawad, N.M.K.; Mansour, D.-E.A. A New Method for Estimating Transformer Health Index Based on Ultraviolet-Visible Spectroscopy. In Proceedings of the 2022 IEEE 23rd International Middle East Power Systems Conference (MEPCON), Cairo, Egypt, 13–15 December 2022; pp. 1–5.
11. Alshehawy, A.M.; Mansour, D.-E.A.; Ghali, M.; Lehtonen, M.; Darwish, M.M.F. Photoluminescence Spectroscopy Measurements for Effective Condition Assessment of Transformer Insulating Oil. *Processes* **2021**, *9*, 732. [CrossRef]
12. Darwish, M.M.F.; Hassan, M.H.A.; Abdel-Gawad, N.M.K.; Mansour, D.-E.A. Application of Infrared Spectroscopy for Discrimination Between Electrical and Thermal Faults in Transformer Oil. In Proceedings of the 2022 9th International Conference on Condition Monitoring and Diagnosis (CMD), Kitakyushu, Japan, 13–18 November 2022; pp. 255–258.
13. *IEC 62770:2013*; Fluids for Electrotechnical Applications—Unused Natural Esters for Transformers and Similar Electrical Equipment, IEC 62770—Edition 1.0. IEC: Geneva, Switzerland, 2013.
14. Redondo EI, D.; Pérez, F.S.; Cantero, M.Á.V.; Zurro, J.I.; Cerdeño JA, R.; Nuñez, J.N. Soil ecotoxicity of natural ester transformer liquids. In Proceedings of the IEEE 18th International Conference on Dielectric Liquids (ICDL), Bled, Slovenia, 29 June–3 July 2014; pp. 1–4.
15. Perrier, C.; Beroual, A.; Bessede, J.-L. Improvement of Power Transformers by using Mixtures of Mineral oil with Synthetic Esters. *J. Colloid Interface Sci.* **2011**, *1*, 71–79.
16. Nadolny, Z.; Dombek, G.; Przybylek, P. Thermal Properties of a Mixture of Mineral Oil and Synthetic Ester in Terms of Its Application in the Transformer. In Proceedings of the IEEE Conference on Electrical Insulation and Dielectric Phenomena (CEIDP), Toronto, ON, Canada, 16–19 October 2016; pp. 857–860.

17. Suwarno; Darma, I.S. Dielectric Properties of Mixtures between Mineral Oil and Natural Ester. In Proceedings of the International Symposium on Electrical Insulating Materials, Yokkaichi, Japan, 7–11 September 2008; pp. 514–517.
18. Hamdi, A.; Fofana, I.; Djillali, M. Stability of mineral oil and oil–ester mixtures under thermal ageing and electrical discharges. *IET Gener. Transm. Distrib.* **2017**, *11*, 2384–2392. [CrossRef]
19. Toudja, T.; Chetibi, F.; Beldjilali, A.; Moulai, H.; Beroual, A. Electrical and physicochemical properties of mineral and vegetable oils mixtures. In Proceedings of the IEEE International Conference on Liquid Dielectrics, Bled, Slovenia, 29 June–3 July 2014; pp. 1–4.
20. Dombek, G.; Gielniak, J. Fire Safety and Electrical Properties of Mixtures of Synthetic Ester/Mineral Oil and Synthetic Ester/Natural Ester. *IEEE Trans. Dielectr. Electr. Insul.* **2018**, *25*, 1846–1852. [CrossRef]
21. Beroual, A.; Sitorus, H.B.H.; Setiabudy, R.; Bismo, S. Comparative Study of AC and DC Breakdown Voltages in Jatropha Methyl Ester Oil, Mineral Oil, and their Mixtures. *IEEE Trans. Dielectr. Electr. Insul.* **2018**, *25*, 1831–1836. [CrossRef]
22. Ghoneim, S.S.M.; Dessouky, S.S.; Boubakeur, A.; Elfaraskoury, A.A.; Sharaf, A.B.A.; Mahmoud, K.; Lehtonen, M.; Darwish, M.M.F. Accurate Insulating Oil Breakdown Voltage Model Associated with Different Barrier Effects. *Processes* **2021**, *9*, 657. [CrossRef]
23. Dixit, A.; Ekanayake, C.; Ma, H.; Saha, T.K.; Ansari, M.H. Thermal Analysis of Natural Cooling Type Distribution Transformer Retrofilled with Natural Ester Oil. *IEEE Trans. Dielectr. Electr. Insul.* **2022**, *29*, 231–239. [CrossRef]
24. Beroual, A.; Khaled, U.; Noah, P.S.M.; Sitorus, H. Comparative Study of Breakdown Voltage of Mineral, Synthetic and Natural Oils and Based Mineral Oil Mixtures under AC and DC Voltages. *Energies* **2017**, *10*, 511. [CrossRef]
25. Das, A.K. Analysis of AC breakdown strength of vegetable oils and effect of mineral oil. *Electr. Power Syst. Res.* **2023**, *214*, 108920. [CrossRef]
26. Alshehawy, A.M.; Mansour, D.A.; Ghali, M. Optical Spectroscopy Analysis of Ester Oils under Thermal Aging Conditions. In Proceedings of the IEEE International Conference on Dielectric Liquids (ICDL), Rome, Italy, 23–27 June 2019; pp. 1–4.
27. Emara, M.M.; Mansour, D.-E.A.; Azmy, A.M. Mitigating the impact of aging byproducts in transformer oil using TiO_2 nanofillers. *IEEE Trans. Dielectr. Electr. Insul.* **2017**, *24*, 3471–3480. [CrossRef]
28. Weibull, W. A statistical distribution function of wide applicability. *ASME J. Appl. Mech.* **1951**, *18*, 293–297. [CrossRef]
29. Mansour, D.A.; Shaalan, E.M.; Ward, S.A.; El Dein, A.Z.; Karaman, H.S.; Ahmed, H.M. Multiple nanoparticles for improvement of thermal and dielectric properties of oil nanofluids. *IET Sci. Meas. Technol.* **2019**, *13*, 968–974. [CrossRef]
30. Mansour, D.A.; Ahmed, H.M.; Salman, A.M. The Effect of Surface Modification of Titania Nanoparticles on the Dielectric Properties of Nanofluids. In Proceedings of the International Middle East Power Systems Conference (MEPCON), Cairo, Egypt, 17–19 December 2019; pp. 1–5.
31. Maher, A.; Mansour, D.A.; Helal, K.; El Aal, R.A.A.A. Dissolved gas analysis and dissipation factor measurement of mineral oil-based nanofluids under thermal and electrical faults. *High Volt.* **2023**, *8*, 455–465. [CrossRef]
32. Farade, R.A.; Wahab, N.I.A.; Mansour, D.-E.A.; Azis, N.B.; Jasni, J.B.; Veerasamy, V.; Vinayagam, A.; Kotiyal, B.M.; Khan, T.M.Y. The Effect of Interfacial Zone Due to Nanoparticle–Surfactant Interaction on Dielectric Properties of Vegetable Oil Based Nanofluids. *IEEE Access* **2021**, *9*, 107033–107045. [CrossRef]
33. Pissis, P.; Fragiadakis, D.; Kanapitsas, A.; Delides, K. Broadband Dielectric Relaxation Spectroscopy in Polymer Nanocomposites. *Macromol. Symp.* **2008**, *265*, 12–20. [CrossRef]
34. Gao, C.; Ye, W.; Xu, Q.; Zhu, M.; Zhu, W.; Hao, J. Influence of Replacing Oil with Different Natural Esters on the Thermal Ageing Behavior of Mineral Oil-Paper Insulation. In Proceedings of the International Conference on Electrical Materials and Power Equipment (ICEMPE), Chongqing, China, 11–15 April 2021; pp. 1–4.
35. Liao, R.J.; Hao, J.; Chen, G.; Ma, Z.Q.; Yang, L.J. A comparative study of physicochemical, dielectric and thermal properties of pressboard insulation impregnated with natural ester and mineral oil. *IEEE Trans. Dielectr. Electr. Insul.* **2011**, *18*, 1626–1637. [CrossRef]
36. Liao, R.; Feng, D.; Hao, J.; Yang, L.; Li, J.; Wang, Q.; Zhang, S. Thermal and electrical properties of a novel 3-element mixed insulation oil for power transformers. *IEEE Trans. Dielectr. Electr. Insul.* **2019**, *26*, 610–617. [CrossRef]
37. Lyutikova, M.; Korobeynikov, S.; Konovalov, A. Evaluation of the Properties of Mixtures of Aromatic Mineral Oil and Synthetic Ester for High-Voltage Equipment. *IEEE Trans. Dielectr. Electr. Insul.* **2021**, *28*, 1282–1290. [CrossRef]
38. Abdel-Gawad, N.M.; El Dein, A.Z.; Mansour, D.A.; Ahmed, H.M.; Darwish, M.M.F.; Lehtonen, M. Development of industrial scale PVC nanocomposites with comprehensive enhancement in dielectric properties. *IET Sci. Meas. Technol.* **2019**, *13*, 90–96. [CrossRef]
39. Abdel-Gawad, N.M.K.; El Dein, A.Z.; Mansour, D.-E.A.; Ahmed, H.M.; Darwish, M.M.F.; Lehtonen, M. Enhancement of dielectric and mechanical properties of Polyvinyl Chloride nanocomposites using functionalized TiO_2 nanoparticles. *IEEE Trans. Dielectr. Electr. Insul.* **2017**, *24*, 3490–3499. [CrossRef]
40. EGYTRAFO. Oil Immersed Transformer Catalogue. Available online: https://egytrafo.com/wp-content/uploads/2020/06/Oil-Transformers-Catalogue.pdf (accessed on 15 July 2021).

41. Bouhaddiche, R.; Bouazabia, S.; Fofana, I. Thermal Modelling of Power Transformer. In Proceedings of the IEEE Conference of Dielectric Liquid (ICDL), Manchester, UK, 25–29 June 2017; pp. 1–4.
42. COMSOL Multiphysics Reference Manual, © 1998–2019 COMSOL, Version: COMSOL 5.5. Available online: https://doc.comsol.com/5.5/doc/com.comsol.help.comsol/COMSOL_ReferenceManual.pdf (accessed on 15 July 2021).

sensors

MDPI

Article

Decentralized Sensor Fault-Tolerant Control of DC Microgrids Using the Attracting Ellipsoid Method

Hisham M. Soliman [1], Ehab H. E. Bayoumi [2], Farag A. El-Sheikhi [3] and Michele De Santis [4,*]

1 Department of Electrical Power Engineering, Faculty of Engineering, Cairo University, Cairo 11562, Egypt; mohammedhisham@eng.cu.edu.eg
2 Department of Mechanical Engineering, Faculty of Engineering, The British University in Egypt, El Sherouk City, Cairo 11837, Egypt; ehab.bayoumi@bue.edu.eg
3 Department of Electrical and Electronics Engineering, Istanbul Esenyurt University, Istanbul 07800, Turkey; farag.elsheikhi@gmail.com
4 Department of engineering, University Niccolò Cusano, 00166 Roma, Italy
* Correspondence: michele.desantis@unicusano.it

Abstract: System stability deterioration in microgrids commonly occurs due to unpredictable faults and equipment malfunctions. Recently, robust control techniques have been used in microgrid systems to address these difficulties. In this paper, for DC-islanded microgrids that have sensors faults, a new passive fault-tolerant control strategy is developed. The suggested approach can be used to maintain system stability in the presence of flaws, such as faulty actuators and sensors, as well as component failures. The suggested control is effective when the fault is never recognized (or when the fault is not being precisely known, and some ambiguity in the fault may be interpreted as uncertainty in the system's dynamics following the fault). The design is built around a derived sufficient condition in the context of linear matrix inequalities (LMIs) and the attractive ellipsoid technique. The ellipsoidal stabilization idea is to bring the state trajectories into a small region including the origin (an ellipsoid with minimum volume) and the trajectories will not leave the ellipsoid for the future time. Finally, computational studies on a DC microgrid system are carried out to assess the effectiveness of the proposed fault-tolerant control approach. When compared with previous studies, the simulation results demonstrate that the proposed control technique can significantly enhance the reliability and efficiency of DC microgrid systems.

Keywords: sensor failure; fault-tolerant control; DC microgrids; attracting ellipsoid method

Citation: Soliman, H.M.; Bayoumi, E.H.E.; El-Sheikhi, F.A.; De Santis, M. Decentralized Sensor Fault-Tolerant Control of DC Microgrids Using the Attracting Ellipsoid Method. *Sensors* **2023**, *23*, 7160. https://doi.org/10.3390/s23167160

Academic Editors: Paweł Rozga, Michał Kunicki and Jan Fulneček

Received: 21 July 2023
Revised: 7 August 2023
Accepted: 11 August 2023
Published: 14 August 2023

1. Introduction

Fault-Tolerant Control (FTC) is concerned with systems whose normal operation is disrupted by a failure in actuators, sensors, or other system components. A component's malfunctioning can be complete, referred to as a failure [1–3], or partial, referred to as fault [4]. This paper focuses on sensor faults.

Passive Fault-Tolerant Control (PFTC) or Active Fault-Tolerant Control (AFTC) can be used to achieve FTC [4]. In the first case, a single controller is designed to stabilize the system and provide the required performance regardless of the failure. This controller does not need to be aware of the fault. AFTC, on the other hand, reconfigures the controller by identifying and isolating the system fault. Many approaches have addressed fault detection and isolation [5].

For example, the sliding-mode observers (SMOs) effectively handle the nonlinear switch discontinuous term for fault detection and estimation. In linear and nonlinear systems with progressive faults, they can handle disturbances, parametric fluctuations, uncertainties, and unmodeled dynamics. It compensates for observer incompatibilities and preserves system stability and reachability in a limited period [6]. Since uncertainties are inherently present, the fault reconstruction technique for matched faults and uncertainties in [7] does not apply to a broad range of actual systems.

Furthermore, they calculated SMO improvements by practically tweaking LMIs according to Lyapunov stability requirements. It lessens the impact of H-∞ criteria, improves fault estimates, and substantiates efforts to address mismatched uncertainty and defects. This study mathematically determined the period the sliding-mode observer would be available and proved its stability. In [8], H-∞ FTC for actuator and sensor faults in wind energy systems is provided. It makes use of Linear-Quadratic Regulator (LQR)-based state feedback control to mimic variable wind speeds generated by stochastic affine models. Event (fault) monitoring for smart grids is provided in [9].

Summarizing: the drawback of AFTC is that it produces a more complex controller. Another disadvantage of AFTC is that fault identification is a challenging task, and in the presence of external disturbances, fault detection algorithms cannot ensure reliable detection. Despite these disadvantages, AFTC has advanced controllers that achieve better performance than PFTC. A PFTC can be expressed as a robust control problem if the sensor's performance is bounded. This method is the focus of this paper.

More renewable energy sources have been integrated into the present power systems as electricity consumption has increased rapidly. This trend has increased the appeal of microgrids (MGs), which provide an effective technique for combining a variety of energy resources (e.g., wind and solar energies) [10–12]. MGs are small-scale power networks comprised of distributed generation (DG), devices for storage, and loads. DC MGs have received increased attention in recent years as a result of the growing use of DC renewables and loads [13–15]. The fundamental goals of DC MG control can be described as voltage regulation, stability, and current sharing. In modern DC MGs, hierarchical control structures are often used to meet these control objectives [16–18].

Control techniques have progressed from (1) centralized where, one controller is used to stabilize the whole system; but requires measuring all the states; (2) decentralized control, where a controller is installed for each DG and requires only the local information of its DG scheme; to (3) distributed schemes, where the controller for each DG requires local and neighboring DGs information [19,20]. It should be noted that the decentralized control is the most reliable option because it does not require the pricey communication network and time delay that centralized control does. Unlike the distributed control which requires communication with neighbor agents, the decentralized control lacks communication. So, this paper focuses on decentralized control.

To improve performance, some hierarchical control techniques have been introduced. In order to track the local voltage reference of DC MGs under plug-and-play (PnP) MG operation, decentralized control method has been presented in [20]. Accurate current sharing has been achieved in [21] by using a secondary consensus-based controller. The typical hierarchical control system, however, has its own limiting conditions because of the complexity of two-layer control of DC MGs [22,23].

Therefore, recent studies [24,25] have considered the constrained communication channel bandwidth, network packet loss, outside interference, and a noisy atmosphere. A distributed secondary cooperative control strategy with adaptive event-triggering communication has been presented in [26]. With reduced communication effort, typical voltage regulation, and appropriate load distribution are achieved. The issue brought on by considering the uncertainty of power loads in DC MGs has been discussed in [27]. To attain satisfactory performance, the authors provided a technique for distributed secondary H∞ consensus. A sliding mode robust controller is incorporated in [28], which studies the impacts of stochastic behavior.

In this paper, a new passive fault-tolerant control strategy is developed for DC-islanded MGs that have voltage sensors' faults. The suggested approach can be used to maintain system stability in the presence of flaws, such as faulty actuators and sensors, as well as component failures. The fault can be unknown to the control system and this ambiguity can be translated as an uncertainty in the dynamics of the system following the fault. The proposed design solves the problem as a robust control problem via a new sufficient linear matrix inequalities (LMI) condition. The invariant ellipsoid method is used to tackle the uncertainty in sensor faults. Finally, the computational results are focused on a DC MG system and are carried out to assess the effectiveness of the proposed fault-tolerant control approach. Compared with previous studies, the simulation results demonstrate that the proposed control technique can significantly enhance the reliability and efficiency of DC MG systems.

The main contributions of this paper are as follows:

(1) A new decentralized voltage tracker design is introduced. The new design technique is based on the Attracting Ellipsoid Approach that is a powerful technique in the robust control theory.
(2) A decentralized state feedback with an integral control is proposed using the current and voltage magnitude of each DG which are the DG states.
(3) To obtain the desired voltage reference tracking performance, it is proposed to use an augmented state feedback controller. Analyzing system stability demonstrates that the suggested controller tolerates sensor faults.
(4) Unlike the difficulties in active fault control (detection and fault evaluation), the proposed robust control is much simpler (one controller), easy to implement, and can cope with sensor fault which is never detected or partially known.

The remainder of this paper is organized as follows: Modeling of DC MG dynamics is briefly discussed in Section 2. Additionally, Section 2 contains the sensor fault model as well as a few more preliminary calculations. Section 3 details the design and analysis of the proposed state feedback with integral control for voltage regulation. In Section 3, the proposed control scheme simulation as well as a comparison with previous published research are provided. Simulation validation of the proposed FTC scheme are detailed in Section 4. Conclusions are provided in Section 5.

2. Problem Formulation and System Modeling

A DC MG comprised of N DGs connected by DC lines is investigated in this paper. Figure 1 depicts the electrical structure of DG-i.

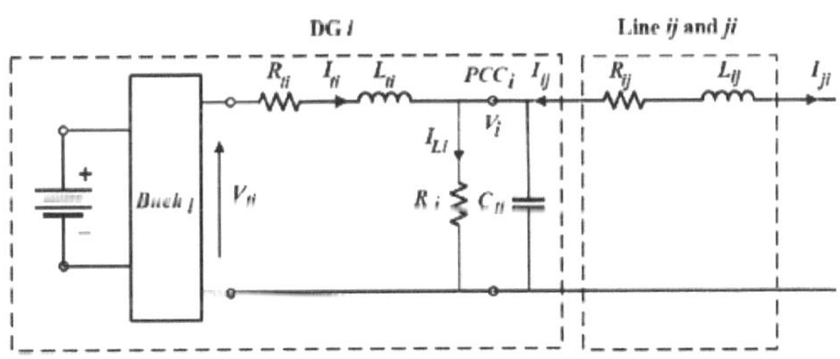

Figure 1. The electrical structure of DG-i.

Each DG has a DC voltage source, a converter, an RLC filter, and a resistive load R_{Li} as shown in Figure 1. It should be noted that the DGs reliability can be affected by the stochastic and intermittent nature of renewable DC energy sources. In practice, there are several options for energy storage systems to solve this problem. In these instances, the sources might be roughly be operated in a steady-state mode. Renewable energy sources are inherently intermittent. In this study, we assume that PVs include a battery storage system to maintain the output voltage constant.

The dynamic model of DG-i is constructed using Kirchhoff's voltage and current laws as follows:

$$\begin{cases} \frac{dV_i}{dt} = \frac{1}{C_{ti}} I_{ti} - \frac{1}{C_{ti}} I_{Li} + \frac{1}{C_{ti}} I_{ij} \\ \frac{dI_{ti}}{dt} = -\frac{1}{L_{ti}} V_i - \frac{R_{ti}}{L_{ti}} I_{ti} + \frac{1}{L_{ti}} V_{ti} \\ Lineij : \frac{dI_{ij}}{dt} = -\frac{R_{ij}}{L_{ij}} I_{ij} + \frac{1}{L_{ij}} V_j - \frac{1}{L_{ij}} V_i \end{cases} \tag{1}$$

where V_i and I_{ti} denote the DG-i capacitor voltage and output current, respectively. The command to the converter is represented by V_{ti}, R_{ti}, L_{ti}, and C_{ti} are constants that represent the filter's electrical properties. R_{ij} and L_{ij} are the power line impedances connecting DG-i and DG-j. Each DG-j's capacitor voltage is represented by V_j.

It is assumed in (1) that the power lines connecting the DGs possess quasi-stationary dynamics [20], i.e., $dI_{ij}/dt = 0$. This assumption is valid as the line inductance L_{ij} in DC systems is significantly small and thereby the line dynamics can be neglected.

$$\begin{cases} \therefore I_{ij} = \frac{V_j - V_i}{R_{ij}} \\ \frac{dI_{ti}}{dt} = -\frac{1}{L_{ti}} V_i - \frac{R_{ti}}{L_{ti}} I_{ti} + \frac{1}{L_{ti}} V_{ti} \\ \frac{dV_i}{dt} = \frac{1}{C_{ti}} I_{ti} - \frac{1}{C_{ti}} I_{Li} + \frac{1}{C_{ti} R_{ij}} V_j - \frac{1}{C_{ti} R_{ij}} V_i \end{cases} \tag{2}$$

The islanded DC MG, shown in Figure 1, contains N DGs that can be modeled through the following state-space equations in the same manner:

$$\dot{x}_i = A_{ii} x_i + B_i u_i + D_i w_i, y_i = C_i x_i = z_i \tag{3}$$

where $x_i = [V_i \; I_{ti}]', u_i = V_{ti}, \; y_i = z_i$ are states vectors, input, y_i as the measurable outputs, and assume that $y_i = x_i$. The output vector to be optimized is z_i. The matrices A_{ij}, B_i, are as follows:

$$\begin{cases} A_{ii} = \begin{bmatrix} -\frac{1}{C_{ti}} \sum_j \frac{1}{R_{ij}} - \frac{1}{R_i C_{ti}} & \frac{1}{C_{ti}} \\ -\frac{1}{L_{ti}} & -\frac{R_{ti}}{L_{ti}} \end{bmatrix} \\ A_{ij} = \begin{bmatrix} \frac{1}{R_{ij} C_{ti}} & 0 \\ 0 & 0 \end{bmatrix} \\ B_i = \begin{bmatrix} 0 \\ \frac{1}{L_{ti}} \end{bmatrix} \\ C_i = \begin{bmatrix} 1 & 0 \\ 0 & 1 \end{bmatrix} \end{cases} \tag{4}$$

The external disturbance is:

$$D_i = [A_{i1} \ldots O_{ii} \ldots A_{iN}], \tag{5}$$

The system studied is chosen to be radial as most of the distribution networks are radial. As illustrated in Figure 2, an islanded DC MG case-study system consists of six DGs.

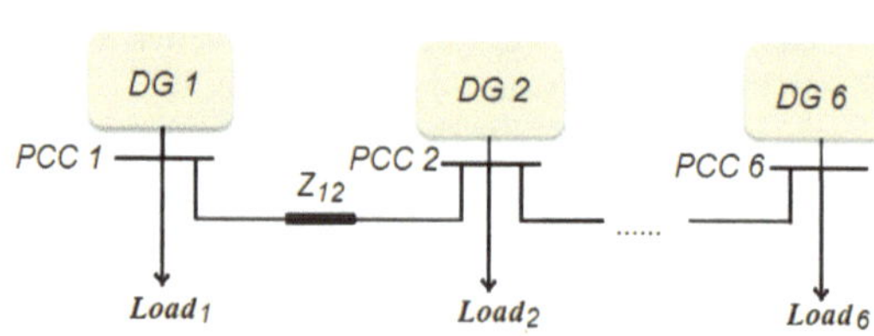

Figure 2. A DC microgrid with six islanded DGs.

Tables 1 and 2 outline the electrical parameters of each distributed generation as well as the distribution lines.

Table 1. Microgrid parameters [20].

DGs	Parameters of the Buck Converter		Shunt Capacitance C_t (mF)	Load Parameter R (Ω)	Power Rating (W)
	R_t (Ω)	L_t (mH)			
DG_1	7.22	72.2	25	160	1200
DG_2	7.22	72.2	32	80	600
DG_3	7.22	72.2	25	120	900
DG_4	7.22	72.2	30	160	1200
DG_5	7.22	72.2	18	100	800
DG_6	7.22	72.2	12	120	900
	V_{dc} (DC bus voltage) 100 V		f_{sw} (Switching frequency) 40 kHz		f_o (Nominal frequency) 50 Hz

Table 2. Parameters of distribution lines.

Line Impedance (Z_{ij})	Line Resistance (R_{ij})			Line Inductance (L_{ij})		
	r_{ij} (Ω/m)	Cable Length (m)	R_{ij} (Ω)	l_{ij} (µH/m)	Cable Length (m)	L_{ij} (µH)
Z_{12}	0.05	180	9	1.8	180	324
Z_{23}	0.05	240	12	1.8	240	432
Z_{34}	0.05	300	15	1.8	300	540
Z_{45}	0.05	240	12	1.8	240	432
Z_{56}	0.05	264	13.2	1.8	264	475.2

The MG study system is discretized with sampling time T_s using Tables 1 and 2. The discrete-time state equation is provided in the Appendix A.

It should be noted that graph theory can be used to solve non-radial networks, as described in [29].

The overall MG discrete-time model is as follows using (3–5):

$$x(k+1) = Ax(k) + Bu(k),\ y(k) = Cx(k),\ x(0) = x_0 \tag{6}$$

The vectors x, u, and y are the state, control, and measurement of dimensions n, m, and l, respectively. Assuming all the states are available for state-feedback control, $C = I$. The control objective aims to obtain the output tracking the input with a steady-state error of zero. In addition, the controller must be decentralized which uses only local states.

The number of outputs that can track a reference input vector, y_r, cannot be more than the number of control inputs to maintain controllability. Consequently, the output equation for the open-loop system shown in (6) can be rewritten as:

$$y(k) = Cx(k) = \begin{bmatrix} C_1 \\ C_2 \end{bmatrix} x(k) = \begin{bmatrix} y_1(k) \\ y_2(k) \end{bmatrix} \tag{7}$$

where $y_1 \in R^h, h \leq l$ denotes the vector of the outputs required to follow the reference input vector y_r. It is to be noticed that the controller is called a regulator if the input is constant; otherwise, it is referred to as a tracker. This section describes the design of the system's decentralized tracker (6). The interconnected system (6) can be subdivided into N subsystems.

With $A = \{A_{i,j}\}$, and $B = blockdiagonl\{B_1, \ldots, B_N\}, C = blockdiagonl\{C_1, \ldots, C_N\}$ subsystem #i is provided by:

$$x_i(k+1) = A_{ii}x_i(k) + B_iu_i(k) + D_ix(k), D_i = \begin{bmatrix} A_{i1} \ldots & O_{ii} \ldots & A_{iN} \end{bmatrix}, y_i(k) = C_ix_i(k) i = 1, .., N \tag{8}$$

The dimensions of x_i, u_i are respectively n_i, m_i, $n = \sum_{i=1}^{N} n_i$, $m = \sum_{i=1}^{N} m_i$.

The decentralization of proposed control can be achieved by reducing the impact of an external disturbance, $D_i\ x(k)$. The dynamics of the remainder of the system on a specific subsystem are represented by Equation (8), where the vector x is supposed to be an external bounded disturbance $w(k)$. The control decentralization can be performed by minimizing the ellipsoid volume, as will be seen in the sequel.

It is worth noting that the MG model (6) does not have an integrator (it is a type 0 plant). As a result, for a step input, a steady-state error will occur. The output voltage must precisely follow the reference voltage with no errors. To achieve tracking task for subsystem i, a vector comparator, and an integrator z_i are added which fulfill:

$$z_i(k+1) = z_i(k) + T_s[y_{ri}(k) - y_{1i}(k)] \tag{9}$$

As a result, the augmented state space representation controls the open-loop system of subsystem-i is:

$$\begin{cases} \hat{x}_i(k+1) = \hat{A}_{ii}\hat{x}(k) + \hat{B}_i u_i(k) + \hat{D}_i w(k) + \hat{I}_i y_{ri}(k), \\ y_i(k) = \hat{C}_i \hat{x}_i(k), \end{cases} \tag{10}$$

where $\hat{x}_i = \begin{bmatrix} x_i \\ z_i \end{bmatrix}, \hat{A}_{ii} = \begin{bmatrix} A_{ii} & O \\ -T_s C_{1i} & I_i \end{bmatrix}, \hat{B}_i = \begin{bmatrix} B_i \\ O \end{bmatrix}, \hat{D}_i = \begin{bmatrix} D_i \\ O \end{bmatrix}, \hat{I}_i = \begin{bmatrix} O \\ T_s I_i \end{bmatrix}, \hat{C}_i = \begin{bmatrix} C_i & 0 \end{bmatrix}.$

It is required to design the state feedback plus integral control provided by:

$$u_i(k) = \hat{K}_i(k)\hat{x}_i(k) = \begin{bmatrix} K_i & K_{Ii} \end{bmatrix} \hat{x}_i(k) \tag{11}$$

Matrix C is an identity matrix for this application.

Now the linear time-invariant dynamical system (10) when subject to sensor faults becomes:

$$\begin{cases} \hat{x}_i(k+1) = \hat{A}_{ii}\hat{x}(k) + \hat{B}_i u_i(k) + \hat{D}_i w(k) + \hat{I}_i y_{ri}(k), \\ y_i(k) = \hat{C}_i \hat{x}_i(k), \\ y_{si}(k) = diag[\varphi_i(k)] y_i(k) \end{cases} \tag{12}$$

where vector $y_i(k)$ represents the system output, and $y_s(k)$ represents the measured output with a sensor fault. The preceding description can be used to model systems with multiplicative faults such as actuator, sensor, and component failures. For modeling the multiplicative faults, the relevant system matrices should be multiplied by the appropriate matrix.

$\varphi_j(k)$ is the sensor function which represents the remaining function of the associated sensor. For example, if a sensor $\varphi_j(k) = 0.8$, in which $\varphi_j(k)$ denotes the remaining function of the jth sensor, then the sensor is 80% functioning. In other words, $\varphi_j(k) = 0$ indicates the sensor failure, while $\varphi_j(k) = 1$ indicates the sensor works properly. A faulted sensor will then be such that $0 < \varphi_j(k) < 1$. Consequently, it is a bounded sensor fault.

The control objective in addition to finding a decentralized dynamic tracker for each DG, should also be robust against sensor fault.

3. Decentralized Passive Sensor Fault-Tolerant Control

Each DG-i is supplied with the proportional with integral controllers listed below, (11):

$$u_i(k) = \hat{K}_i(k)\hat{x}_i(k) = \begin{bmatrix} K_i & K_{Ii} \end{bmatrix} \hat{x}_i(k) \tag{13}$$

where $K_i \in R^{1\times 2}$, $K_{Ii} \in R^{1\times 1}$, and controllers, $i = 1, \ldots, N$, are decentralized because the u_i computation only requires the DG-i state. The proposed system incorporates proportional state feedback and integral control. It is completely decentralized by means of local states. It avoids communication of bordering subsystems states as in the case of distributed control. The proposed design is simple, contrary to centralized control, which has an expensive communication network and associated delay, which reduces system stability. It should be noted that the communication network in a centralized scheme is prone to failure, which can lead to a total collapse of the control system.

The invariant-set approach [30,31], and its reference applications [32–34], is an approach used for designing systems controllers that are disrupted by external disturbances (perturbed systems). An invariant-set is one in which if an initial state vector $x(0)$ begins within it, the trajectory $x(k)$ will not leave it for the future time $k > 0$. The invariant set is approximated using a bounding ellipsoid because determining it mathematically is difficult. The invariant ellipsoid technique is employed in linear systems to suppress bounded disturbance by decreasing the ellipsoid volume. In [30], the invariant ellipsoid approach is employed to propose a novel strategy for minimizing the effect of external disturbances on linear systems. When the initial state is outside of the ellipsoid, the ellipsoid is a set that contains the origin and attracts the state trajectory. As a result, it is referred to as an attractive ellipsoid. When the state trajectory arrives at the ellipsoid, it does not leave it as time passes. As a result, the ellipsoid is referred to as an invariant ellipsoid. To reduce the effect of external disturbances on the trajectory, the volume of the ellipsoid must be reduced [31]. The goal of the MG voltage control challenge is to develop a controller that allows the output voltage to track the reference voltage.

The tracker must also be long-lasting in the event of sensor faults. It should also lessen the impact of disruptions on the output voltage. This is known as a disturbance-rejection tracker.

In (10), $w(k)$ reflects the bounded external disturbances that are subject to the constraint:

$$| \, | w(k) | \, | \leq 1, \forall k \geq 0 \tag{14}$$

The symbol $||(.)||$ denotes the vector (.) Euclidean norm. Note that, the external disturbance is L∞-bounded. To optimize z, the design goal is also to minimize the influence of disturbance Dw on the output. It should be observed that the disturbance constraint (14) has no effect on generality because the matrix D can always be scaled to satisfy (14). Note that the normalization in constraint (14) results in simpler LMI condition than if normalization is not carried out.

Ref. [30] considers the following problem. Given the discrete time system.

$$x(k+1) = Ax(k) + Bu(k) + Dw(k),\ y(k) = Cx(k),\ subject\ to\ \|w(k)\| \leq 1 \tag{15}$$

The pairs (A, B) and (A, C) are assumed controllable, and observable respectively. The state feedback controller $u(k) = Kx(k)$, which stabilizes (14) and rejects the disturbance $w(k)$ in an ideal way (in terms of minimizing the bounding ellipsoid trace of the optimized output, $E_z = CPC\prime$) is provided by the following theorem [30].

Notation. *The superscript $(.)'$ represents matrix transposition throughout the paper, R^n denotes the n-dimensional Euclidean space and $R^{n\times m}$ is the set of all $n \times m$ real matrices. For a symmetric $P \in R^{n\times n}$, $P > 0$ indicates that it is positive definite. A symmetric matrix $\begin{bmatrix} Q+Z+Q'+Z' & R \\ R\prime & P \end{bmatrix}$ is denoted by $\begin{bmatrix} (Q+Z+*) & R \\ * & P \end{bmatrix}$*

Theorem 1. *Ref. [30]*

Let P, Y be a solution of the minimization optimization problem.

$$minimize tr CPC\prime$$

Subject to the constraints:

$$\begin{bmatrix} -\alpha P & * & * \\ AP+BY & -P & * \\ 0 & D' & -(1-\alpha)I \end{bmatrix} \leq 0,\ P > 0,\ \alpha > 0$$

For some $0 < \alpha < 1$. The minimization is carried out with respect to the matrix variables $P = P'$, Y, and the scalar parameter α.

Moreover, the optimal state controller, stabilizing (15) and rejecting the disturbance, is provided by:

$$K = YP^{-1}$$

Note that the trace function is adopted due to its linearity; being synonymous to the sum of squared semi-axes of the ellipsoid E_z. This latter condition will be employed throughout the text to help reduce the problem of interest to standard semidefinite programs. The multiplicative term αP is the source of nonlinearity in the above theorem. After fixing α, the above matrix equation becomes linear, making it simple to solve with the LMI toolbox. The scalar α is iteratively updated to solve the minimization problem.

Theorem 1 can be used to solve the sensor FTC as follows. After the occurrence of the sensor fault, the matrix C_i in (4) becomes one of the matrices $diag\left[\varphi_{i,j}(t)\right]C_i, i = 1\ldots N$, $j = 1,2$.

The effectiveness reduction in the sensors of DG-i is randomly selected as $\varphi_{i,1} = diag[0.81\ 0.91]$, $\varphi_{i,2} = diag[0.91\ 0.63]$.

The design of sensor FTC for DG-i can be performed by replacing C in Theorem 1, by $diag\left[\varphi_{i,j}(t)\right]C_i$, A by $\hat{A}_i$, and B by $\hat{B}_i$. We obtain the following theorem.

Theorem 2. *The sensor FTC of DG-i, i = 1, . . ., N, is obtained by solving the following optimization problem:*

$$minimize\left\{max_{j=1,2}tr\left\{\left(diag\left[\varphi_{i,j}(t)\right]C_i\right)P_i\left(diag\left[\varphi_{i,j}(t)\right]C_i\right)'\right\},\ i = 1\ldots N\right\}.$$

Subject to the constraints:

$$\begin{bmatrix} -\alpha\hat{P}_i & * & * \\ \hat{A}_i\hat{P}_i + \hat{B}_i\hat{Y}_i & -\hat{P}_i & * \\ 0 & D\prime & -(1-\alpha)I \end{bmatrix} \leq 0,\ \hat{P}_i > 0, \alpha > 0$$

Moreover, the optimal sensor FTC is provided by:

$$\hat{K}_i = \hat{Y}_i\hat{P}_i^{-1},\ i = 1\ldots N$$

Solving Theorem 2 (using Matlab LMI, yalmip, and sedumi), the proposed tracker is provided in Table 3.

Table 3. The proposed tracker.

Controller	α	Ellipsoid Volume	K, K_I
1	0.05	1.6509×10^{-16}	[−14.649 −45.078], 42.592
2	0.42	1.639×10^{-16}	[−59.741 −74.988], 155.51
3	0.03	2.5792×10^{-16}	[−24.598 −55.922], 22.399
4	0.1	2.6294×10^{-16}	[−18.248 −43.158], 52.527
5	0.16	2.8791×10^{-16}	[−98.091 −114.53], 108.91
6	0.42	5.9235×10^{-16}	[−85.036 −86.043], 145.12

4. Simulation Validation

The system shown in Figure 2 is modelled using the Matlab/SimPower Systems Toolbox. Robust stability, required response, and steady-state capabilities have all been achieved as per the IEEE requirements [35].

The performance of the developed controllers is measured during the randomized effectiveness of the sensor signal (sensor fault). The suggested planned controls are tested in four different scenarios on the study system. Each of the four scenarios is played out by randomly selecting a sensor in a certain DG that has a failure level (reduction in the effectiveness of the sensor signal in percentage).

Table 4 shows the random sensor faults choices applied to random DGs. The introduced design is a proportional state-feedback system with integral control that is totally decentralized by utilizing local states. Therefore, six controllers are designed, and their gains are provided in Table 3. Note that Theorem 2 is only a sufficient condition so the proposed controller is stabilizing the system for faults more severe than the design faults, Table 4.

Table 4. Reduction in the effectiveness of the sensor signal in percentage.

Scenario Number	Case	DG Number	Time (s)	Effectiveness of the Sensor Signal
1	Case 1 Case 2	DG_1 DG_5	at t = 7 s at t = 8 s	[0.8 0.6] [0.9 0.7]
2	One case only	Consecutive faults on DG_2 and DG_4	at t = 8 s at t = 9 s	[0.8 0.65] (DG_2) [0.9 0.75] (DG_4)
3	One case only	Simultaneous faults on DG_2 and DG_4 (within the design control range)	at t = 9 s and at t = 9 s	[0.8 0.65] (DG_2) [0.9 0.75] (DG_4)
4	One case only	Simultaneous faults on DG_2 and DG_4 (outside the design control range)	at t = 9 s and at t = 9 s	[0.6 0.55] (DG_2) [0.5 0.35] (DG_4)

To make a comparative study between the proposed technique and other techniques, a completely decentralized auto-tuned control method is used to design six PI controllers for the six DGs provided in Figure 2. The designed gains for the six auto-tuned PI controllers are illustrated in Table 5.

Table 5. Auto-tuned PI controllers gains for the six DGs.

PI-Gains	DG_1	DG_2	DG_3	DG_4	DG_5	DG_6
K_p	2.623101	1.554792	2.095479	1.958416	2.518254	2.428056
K_i	50.52526	23.36601	21.52614	21.11258	30.13315	54.55977

Remark 1. *Summary of the proposed and auto-tuned control algorithms.*

The proposed algorithm:

- For a given scalar α, the matrix equations in Theorem 2 become linear, solves them by the Matlab LMI toolbox.
- Calculates the objective function $\{\max_{j=1,2} tr\{(diag[\varphi_{i,j}(t)]C_i)P_i(diag[\varphi_{i,j}(t)]C_i)'\}, i = 1 \ldots N\}$.
- Updates α iteratively till the minimum of the objective function is obtained (the Matlab command fminsearch can be used).

The auto-tuned algorithm:

The method of setting controller gains based on a study system model or data is known as auto-tune PI. It tunes PI gains in a Simulink model using Simulink Control Design^{TM}. The auto-tune PI controller operates using a linearization of the study system model. It computes PI controller gains based on the obtained response to balance robustness and performance.

4.1. Scenario 1: The Sensor's Effectiveness Reduction in one DG

4.1.1. Case 1: Sensor Failure in DG_1

By constructing a random selection approach and picking a problem in one sensor at a random deterioration level at a random time, we were able to choose the sensor in DG_1 with 80% sensor effectiveness at t = 7 s, as shown in Table 4.

At t = 7 s, Figure 3 shows the DGs voltage during sensor effectiveness in DG_1 is degraded from 100% to 80%: Figure 3a,c for the conventional PI auto-tuned technique, and Figure 3b,d for the proposed control technique.

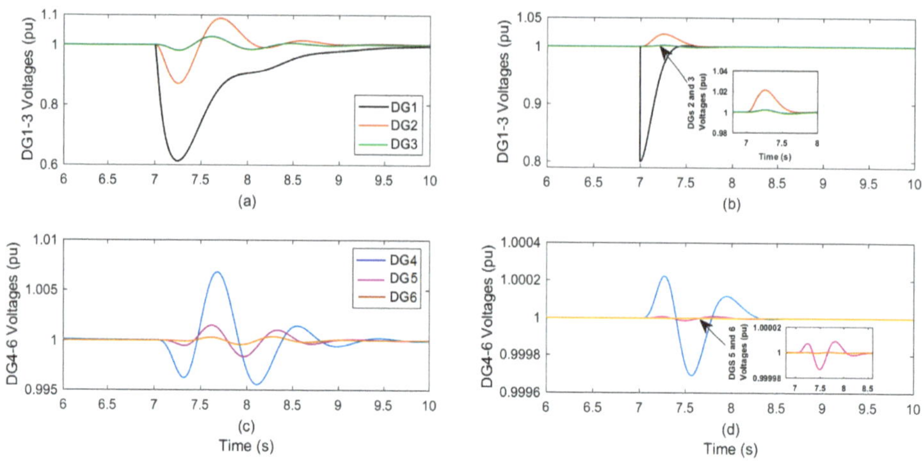

Figure 3. The DGs voltage during sensor effectiveness in DG1 is degraded from 100% to 80%; (**a**,**c**) the conventional PI-tuned; (**b**,**d**) the proposed control.

The fault impacts in the other five DGs are significantly severe in the auto-tuned approach while, it is minimal in the proposed one, the effects are close to zero. Table 6 summarizes the implications of 80% sensor effectiveness in DG_1 on the other five DGs for both control techniques.

Table 6. Controller response parameters during sensor effectiveness fault 80% in DG_1.

		% of Voltage Dip during Sensor Fault (%)		Control Response Parameters					
				% Overshoot (%)		Settling Time (s)		Steady State Error (%)	
		PI-Tuned	Proposed	PI-Tuned	Proposed	PI-Tuned	Proposed	PI-Tuned	Proposed
Output Voltage	DG_1	39.12	19.35	≈0.0	≈0.0	2.312	0.482	0.012	≈0.0
	DG_2	12.32	0.252	9.843	2.871	1.834	0.435	≈0.0	≈0.0
	DG_3	4.781	0.323	2.351	0.335	1.237	0.351	≈0.0	≈0.0
	DG_4	0.431	0.031	0.734	0.0213	1.051	0.336	≈0.0	≈0.0
	DG_5	0.219	0.0017	0.204	0.00123	1.047	0.271	≈0.0	≈0.0
	DG_6	0.078	0.00042	0.197	≈0.0	1.039	0.123	≈0.0	≈0.0

The control parameters shown in Table 6 demonstrate the dead-beat, quick, and zero steady-state performance of the proposed designed trackers for the six DGs and the auto-tuned PI at 80% sensor effectiveness in DG_1.

4.1.2. Sensor Failure in DG_5

As shown in Table 4, selecting the sensor in DG_5 with 90% sensor effectiveness at t = 8 s using a random selection technique and a sensor fault at a random degradation level and duration. At t = 8 s the voltage of the DGs is shown in Figure 4a,c for the traditional

PI auto-tuned approach, and in Figure 4b,d for the suggested control technique, at a time when the sensor effectiveness in DG5 decreased from 100% to 90%.

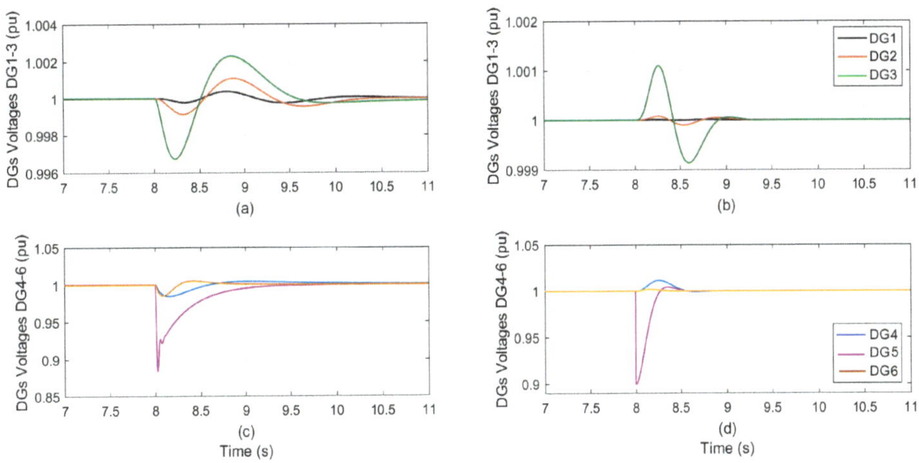

Figure 4. The DGs voltage during sensor effectiveness in DG_5 is degraded from 100% to 90%; (**a**,**c**) the conventional PI-tuned; (**b**,**d**) the proposed control.

The proposed technique has only slight effects on the other five DGs compared with the auto-tuned method, which has a significant impact not only on the faulty sensor DG but also on the remaining five DGs. Table 7 provides a concise summary of the impact that each control technique had on the faulty sensor DG as well as the other five DGs.

Table 7. Controller response parameters during sensor effectiveness fault 90% in DG_5.

		% of Voltage Dip during Sensor Fault (%)		Control Response Parameters					
				% Overshoot (%)		Settling Time (s)		Steady State Error (%)	
		PI-Tuned	Proposed	PI-Tuned	Proposed	PI-Tuned	Proposed	PI-Tuned	Proposed
Output Voltage	DG_1	0.043	0.0044	0.047	≈0.0	1.037	0.311	0.0005	≈0.0
	DG_2	0.106	0.013	0.148	0.035	1.049	0.413	0.0008	≈0.0
	DG_3	0.237	0.098	0.234	0.124	1.121	0.532	0.0009	≈0.0
	DG_4	2.232	0.217	1.12	1.45	1.534	0.456	0.047	≈0.0
	DG_5	12.07	9.841	≈0.0	0.129	1.765	0.512	0.051	≈0.0
	DG_6	2.354	0.221	1.17	0.892	1.627	0.488	0.0481	≈0.0

4.2. Scenario 2: Successive Sensor's Effectiveness Degrades in Two DGs

The sensor in DG_2 was randomly picked with 80% effectiveness at t = 8 s while the sensor in DG_4 was randomly selected with 90% effectiveness at t = 9 s using a randomized selection approach, as shown in Table 4.

Figure 5a,c depict the two successive sensor failures in DG_2 and DG_4 when the conventional auto-tuned PI control method is utilized. While Figure 5b,d depict two successive sensor errors in DG_2 and DG_4 when the proposed control method is operating.

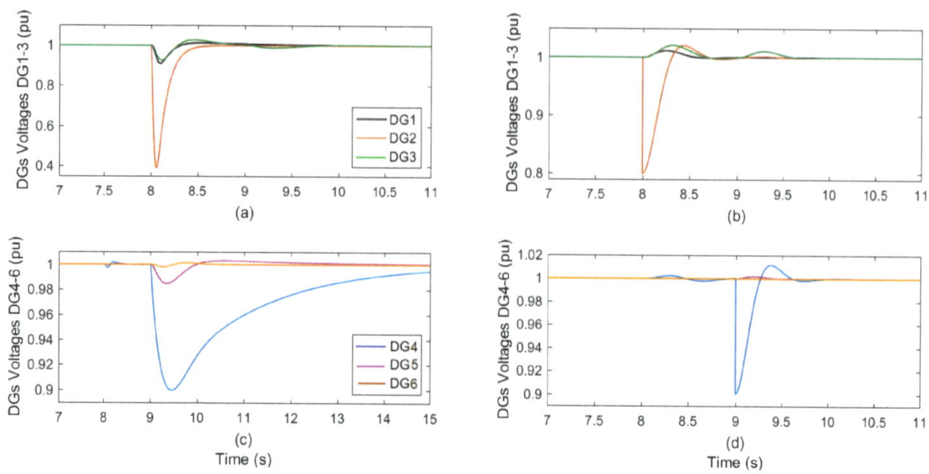

Figure 5. The DGs voltage during successive sensor's effectiveness degrades in DG_2 (80%) followed by DG_4 (90%); (**a**,**c**) the conventional PI-tuned; (**b**,**d**) the proposed control.

Since the proposed tracker approaches the fault problem as a disturbance, the fault effects arising in neighboring DG_1 and DG_3 during the first fault occurrence in DG_2 are minimal, as shown in Figure 5b. Moreover, the proposed controls performed admirably during the successive sensor failure in DG_4 and its neighboring DGs.

Figure 5a,c show the effect on the faulty sensor DGs and their neighboring DGs when the auto-tuned PI controllers are operating.

Table 8 presents a condensed overview of the influence that both control techniques made on the six DGs caused by simultaneous faults in the sensors of DG_2 and DG_4.

Table 8. Controller response parameters during simultaneous 80% sensor effectiveness in DG_2 followed by 90% sensor effectiveness in DG_4.

		% of Voltage Dip during Sensor Fault (%)		Control Response Parameters					
				% Overshoot (%)		Settling Time (s)		Steady State Error (%)	
		PI-Tuned	Proposed	PI-Tuned	Proposed	PI-Tuned	Proposed	PI-Tuned	Proposed
		DG_2	DG_4	DG_2 DG_4	DG_2 DG_4	DG_2 DG_4	DG_2 DG_4	DG_2 DG_4	DG_2 DG_4
Output Voltage	DG_1	9.743	0.031	3.85 1.03	1.65 0.15	1.52 1.77	0.51 0.17	0.08 0.02	≈0.0 ≈0.0
	DG_2	60.972	19.972	≈0.0 1.45	2.47 0.37	1.67 1.96	0.57 0.37	0.09 0.08	≈0.0 ≈0.0
	DG_3	9.534	0.0287	2.13 3.27	2.78 1.34	1.48 2.34	0.49 0.49	0.03 0.47	≈0.0 ≈0.0
	DG_4	10.892	9.985	0.89 ≈0.0	0.51 1.73	1.29 5.92	0.31 0.52	≈0.0 1.89	≈0.0 ≈0.0
	DG_5	1.842	0.0224	0.72 1.13	0.37 0.46	1.15 3.46	0.26 0.34	≈0.0 0.69	≈0.0 ≈0.0
	DG_6	0.351	0.0169	0.19 0.78	0.03 0.23	1.07 1.78	0.19 0.24	≈0.0 0.35	≈0.0 ≈0.0

4.3. Scenario 3: Concurrent Sensor's Effectiveness Degrades in Two DGs inside the Designed Range

Using randomized selection, the sensor in DG_2 was chosen at random with an effectiveness of 80% at t = 9 s, while the sensor in DG_4 was selected at random with an effectiveness of 90% at the same time, as shown in Table 4.

When the traditional auto-tuned PI control approach is applied, both DG_2 and DG_4 have concurrent sensor failures, which are shown in Figure 6a,c. Figure 6b,d show two different sensor faults happening at the same time in DG_2 and DG_4 while the suggested control technique is being applied.

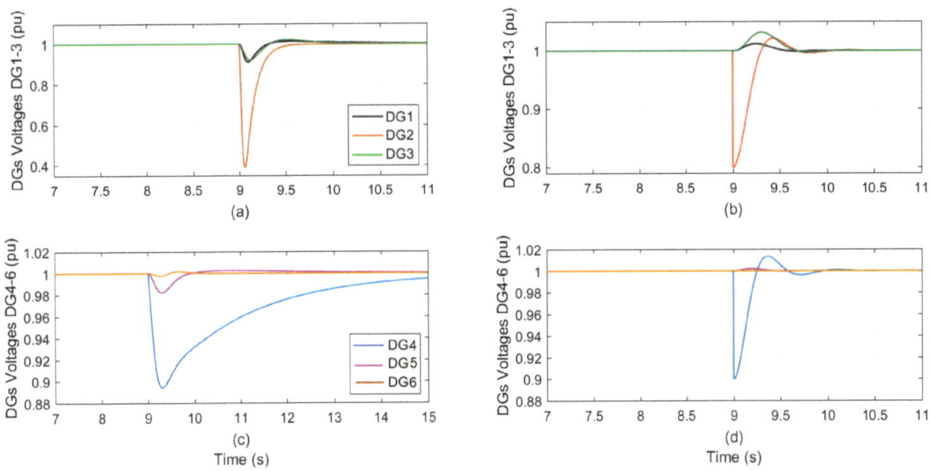

Figure 6. The DGs voltage during concurrent sensor's effectiveness degrades in DG_2 (80%) and in DG_4 (90%) concurrently; (**a**,**c**) the conventional PI-tuned; (**b**,**d**) the proposed control.

In Figure 6b,d the proposed controllers are tested against this scenario. The results are much better than those for the conventional auto-tune PI controllers provided in Figure 6a,c.

Concurrent faults in the sensors of DG_2 and DG_4 are summarized in Table 9, which provides an overview of the impact of both control strategies on all six DGs.

Table 9. Controller response parameters during concurrent 80% sensor effectiveness in DG_2 and 90% sensor effectiveness in DG_4 inside the designed range.

		% of Voltage Dip during Sensor Fault (%)		Control Response Parameters					
				% Overshoot (%)		Settling Time (s)		Steady State Error (%)	
		PI-Tuned	Proposed	PI-Tuned	Proposed	PI-Tuned	Proposed	PI-Tuned	Proposed
Output Voltage	DG_1	8.729	≈0.0	0.4367	3.387	0.9219	0.6287	0.00781	≈0.0
	DG_2	61.251	19.871	≈0.0	2.031	0.9512	0.6529	0.1034	≈0.0
	DG_3	8.5625	≈0.0	0.4298	1.436	0.9037	0.6194	0.00651	≈0.0
	DG_4	10.389	9.934	≈0.0	1.865	4.672	0.468	0.6621	≈0.0
	DG_5	1.9761	≈0.0	0.311	0.461	3.473	0.211	0.1036	≈0.0
	DG_6	0.8747	≈0.0	0.2984	0.207	1.267	0.1057	0.01453	≈0.0

4.4. Scenario 4: Concurrent Sensor's Effectiveness Degrades in Two DGs outside the Designed Range

To put the proposed system through rigorous testing. The suggested system is subjected to two DG sensors operating concurrently with effectiveness values beyond the planned control range.

This was accomplished by picking DG_2 and DG_4 at random, with sensor failure effectiveness as shown in Table 4.

The proposed system operated perfectly and rejected the sensors' disturbance successfully. The proposed trackers respond swiftly and sensor's fault reflection to the neighboring DGs is minimal, as shown in Figure 7a,b.

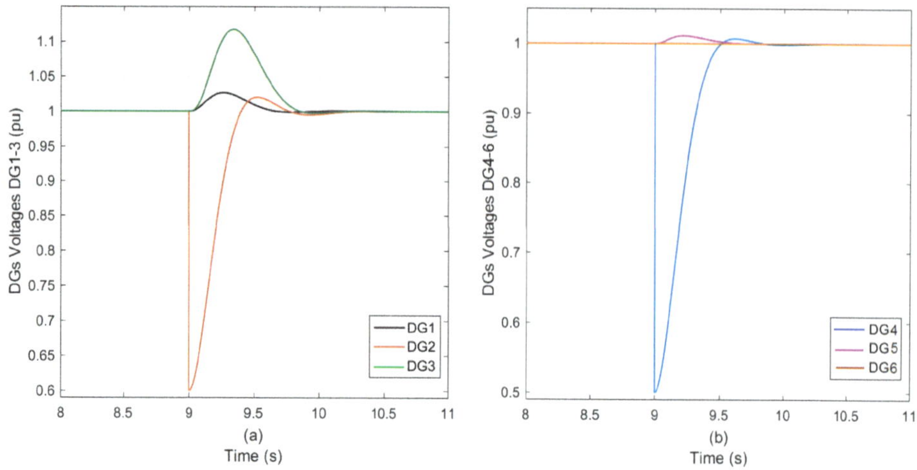

Figure 7. The DGs voltage during concurrent sensor's effectiveness degrades in DG_2 (60%) followed by DG_4 (50%) outside the design range using the proposed control; (**a**) DG_1, DG_2, and DG_3; (**b**) DG_4, DG_5, and DG_6.

The control response parameters for the six DGs are illustrated in Table 10.

Table 10. Controller response parameters during sensor effectiveness fault 90% in DG_5.

		% of Voltage Dip during Sensor Fault (%)	Control Response Parameters		
			% Overshoot (%)	Settling Time (s)	Steady State Error (%)
		Proposed	Proposed	Proposed	Proposed
Output Voltage	DG_1	Proposed	Proposed	Proposed	Proposed
	DG_2	≈0.0	3.1046	0.6387	≈0.0
	DG_3	39.8734	2.2641	0.7793	≈0.0
	DG_4	≈0.0	12.037	0.8367	≈0.0
		49.5439	1.3474	0.8169	≈0.0
	DG_5	≈0.0	2.0413	0.4536	≈0.0
	DG_6	≈0.0	0.2281	0.1106	≈0.0

5. Conclusions

The issue of FTC for DC microgrids is investigated in this paper. The introduced technique for mitigating the impact of sensor faults is a passive FTC scheme. By considering the microgrid under random sensors' faults, conditions are obtained for PFTC to achieve stability of the closed loop.

These conditions are derived in terms of LMIs for the proposed state feedback with an integral voltage tracker. The results are obtained via modeling the sensors fault as a norm-bounded type parameter uncertainty. The effects of such uncertainties on the system performance are attenuated by minimizing the relevant attracting ellipsoid.

Results from the analysis and simulation studies reveal that the proposed controller has satisfied performance even with the simultaneous faulty sensors and measurements. DC microgrids are susceptible to a variety of failures and faults in practical applications.

Future research should investigate the impact of actuator faults and other types of fault signals on the functioning of DC microgrids. In addition, it should be investigated if the sensor FTC be viewed from another perspective (as if the system is under cyber-attack [36]). Further study is needed in this direction.

Author Contributions: Conceptualization, H.M.S. and E.H.E.B.; methodology, E.H.E.B., F.A.E.-S. and H.M.S.; software, H.M.S., F.A.E.-S. and E.H.E.B.; validation, M.D.S., E.H.E.B., F.A.E.-S. and H.M.S.; formal analysis, M.D.S., E.H.E.B., F.A.E.-S. and H.M.S.; investigation, M.D.S., E.H.E.B., F.A.E.-S. and H.M.S.; resources, H.M.S., E.H.E.B. and M.D.S.; data curation, M.D.S., E.H.E.B., F.A.E.-S. and H.M.S.; writing—original draft preparation, M.D.S., E.H.E.B., F.A.E.-S. and H.M.S.; writing—review and editing, M.D.S., E.H.E.B., F.A.E.-S. and H.M.S.; visualization, M.D.S., E.H.E.B., F.A.E.-S. and H.M.S.; supervision, M.D.S., E.H.E.B., F.A.E.-S. and H.M.S. All authors have read and agreed to the published version of the manuscript.

Funding: This research received no external funding.

Institutional Review Board Statement: Not applicable.

Informed Consent Statement: Not applicable.

Data Availability Statement: Not applicable.

Conflicts of Interest: The authors declare no conflict of interest.

Appendix A

The DC microgrid's state equation matrices:

$$A = \begin{bmatrix}
0.9951 & 0.03797 & 0.004429 & 6.703 \times 10^{-5} & 5.77 \times 10^{-6} & 7.51 \times 10^{-8} \\
-0.01315 & 0.9046 & -2.971 \times 10^{-5} & -3.045 \times 10^{-7} & -2.601 \times 10^{-8} & -2.563 \times 10^{-105} \\
0.004429 & 8.58 \times 10^{-5} & 0.998 & 0.0297 & 0.002599 & 5.032 \times 10^{-5} \\
-1.49 \times 10^{-5} & -1.954 \times 10^{-7} & -0.006601 & 0.9045 & -8.736 \times 10^{-6} & -1.146 \times 10^{-7} \\
5.77 \times 10^{-6} & 7.511 \times 10^{-8} & 0.002599 & 3.931 \times 10^{-5} & 0.998 & 0.03802 \\
-1.738 \times 10^{-8} & -1.713 \times 10^{-10} & -1.165 \times 10^{-5} & -1.194 \times 10^{-7} & -0.008802 & 0.9044 \\
5.131 \times 10^{-9} & 5.033 \times 10^{-11} & 3.466 \times 10^{-6} & 3.523 \times 10^{-8} & 0.002662 & 5.153 \times 10^{-5} \\
-1.743 \times 10^{-11} & -1.376 \times 10^{-13} & -1.562 \times 10^{-8} & -1.202 \times 10^{-10} & -1.785 \times 10^{-5} & -2.341 \times 10^{-7} \\
3.563 \times 10^{-12} & 2.805 \times 10^{-14} & 3.208 \times 10^{-9} & 2.458 \times 10^{-11} & 3.696 \times 10^{-6} & 4.809 \times 10^{-8}8 \\
-4.869 \times 10^{-15} & -3.209 \times 10^{-17} & -5.463 \times 10^{-12} & -3.37 \times 10^{-14} & -8.35 \times 10^{-9} & -8.228 \times 10^{-11} \\
2.998 \times 10^{-15} & 1.971 \times 10^{-17} & 3.374 \times 10^{-12} & 2.075 \times 10^{-14} & 5.181 \times 10^{-95} & 5.082 \times 10^{-11} \\
-4.563 \times 10^{-18} & -2.581 \times 10^{-20} & -6.148 \times 10^{-15} & -3.165 \times 10^{-17} & -1.176 \times 10^{-11} & -9.29 \times 10^{-14}
\end{bmatrix}$$

$$\begin{bmatrix}
5.131 \times 10^{-9} & 4.194 \times 10^{-11} & 3.563 \times 10^{-12} & 3.896 \times 10^{-14} & 2.998 \times 10^{-15} & 4.107 \times 10^{-17} \\
-1.743 \times 10^{-11} & -1.147 \times 10^{-13} & -9.712 \times 10^{-15} & -8.889 \times 10^{-17} & -6.826 \times 10^{-18} & -8.042 \times 10^{-20} \\
3.466 \times 10^{-6} & 3.758 \times 10^{-8} & 3.208 \times 10^{-9} & 4.37 \times 10^{-11} & 3.374 \times 10^{-12} & 5.533 \times 10^{-14} \\
-7.83 \times 10^{-9} & -6.429 \times 10^{-11} & -5.463 \times 10^{-12} & -5.991 \times 10^{-14} & -4.611 \times 10^{-15} & -6.33 \times 10^{-17} \\
0.002662 & 4.295 \times 10^{-5} & 3.696 \times 10^{-6} & 6.68 \times 10^{-8} & 5.181 \times 10^{-9} & 1.059 \times 10^{-10} \\
-1.193 \times 10^{-5} & -1.304 \times 10^{-7} & -1.113 \times 10^{-8} & -1.524 \times 10^{-10} & -1.176 \times 10^{-11} & -1.935 \times 10^{-13}3 \\
0.9983 & 0.0317 & 0.002772 & 7.454 \times 10^{-5} & 5.827 \times 10^{-6} & 1.58 \times 10^{-7} \\
-0.0131 & 0.9046 & -1.858 \times 10^{-5} & -3.386 \times 10^{-7} & -2.626 \times 10^{-8} & -5.393 \times 10^{-10}0 \\
0.002772 & 4.472 \times 10^{-5} & 0.9971 & 0.05279 & 0.004192 & 0.0001692 \\
-9.318 \times 10^{-6} & -1.019 \times 10^{-7} & -0.006599 & 0.9047 & -1.41 \times 10^{-5} & -3.854 \times 10^{-7} \\
5.827 \times 10^{-6} & 6.321 \times 10^{-8} & 0.004192 & 0.0001128 & 0.9947 & 0.07908 \\
-1.756 \times 10^{-8} & -1.442 \times 10^{-10} & -1.88 \times 10^{-5} & -3.426 \times 10^{-7} & -0.008787 & 0.9042
\end{bmatrix}$$

$$B = \begin{bmatrix} 0.0002676 & 0.0 & 0.0 & 0.0 & 0.0 & 0.0 \\ 0.01318 & 0.0 & 0.0 & 0.0 & 0.0 & 0.0 \\ 0.0 & 0.0001049 & 0.0 & 0.0 & 0.0 & 0.0 \\ 0.0 & 0.006607 & 0.0 & 0.0 & 0.0 & 0.0 \\ 0.0 & 0.0 & 0.000179 & 0.0 & 0.0 & 0.0 \\ 0.0 & 0.0 & 0.00881 & 0.0 & 0.0 & 0.0 \\ 0.0 & 0.0 & 0.0 & 0.0002232 & 0.0 & 0.0 \\ 0.0 & 0.0 & 0.0 & 0.01318 & 0.0 & 0.0 \\ 0.0 & 0.0 & 0.0 & 0.0 & 0.0001865 & 0.0 \\ 0.0 & 0.0 & 0.0 & 0.0 & 0.006608 & 0.0 \\ 0.0 & 0.0 & 0.0 & 0.0 & 0.0 & 0.0003726 \\ 0.0 & 0.0 & 0.0 & 0.0 & 0.0 & 0.008809 \end{bmatrix}$$

C is a unit matrix.

References

1. El-Sheikhi, F.A.; Soliman, H.M.; Ahshan, R.; Hossain, E. Regional Pole Placers of Power Systems under Random Failures/Repair Markov Jumps. *Energies* **2021**, *14*, 1989. [CrossRef]
2. Soliman, H.M.; Ghommam, J. Reliable control of power systems. In *Diagnosis, Fault Detection & Tolerant Control*; Springer: Berlin/Heidelberg, Germany, 2020. [CrossRef]
3. Bayoumi, E.H.E.; Soliman, H.M.; Albadi, M.; Soliman, M. Invariant set design of decentralized control for islanded microgrids under plug-and-play operation. *Int. J. Electr. Power Energy Syst.* **2021**, *128*, 106678. [CrossRef]
4. Gershon, E.; Shaked, U. Robust Switching-Based Fault-Tolerant Control. In *Advances in H∞ Control Theory: Switched, Delayed, and Biological Systems*; Springer: Berlin/Heidelberg, Germany, 2019.
5. Mahmoud, M.S.; Xia, Y. *Analysis and Synthesis of Fault-Tolerant Control Systems*; John Wiley & Sons, Ltd.: Hoboken, NJ, USA, 2014.
6. Alwi, H.; Edwards, C.; Marcos, A. Fault reconstruction using a LPV sliding mode observer for a class of LPV systems. *J. Frankl. Inst.* **2012**, *349*, 510–530. [CrossRef]
7. Dhahri, S.; Sellami, A.; Hmida, F.B. Robust sliding mode observer design for a class of uncertain linear systems with fault reconstruction synthesis. *Int. J. Phys. Sci.* **2012**, *7*, 1259–1269. [CrossRef]
8. Shi, Y.-T.; Kou, Q.; Sun, D.-H.; Li, Z.-X.; Qiao, S.-J.; Hou, Y.-J. H_∞ Fault Tolerant Control of WECS Based on the PWA Model. *Energies* **2014**, *7*, 1750–1769. [CrossRef]
9. Faheem, M.; Butt, R.A.; Raza, B.; Ashraf, M.W.; Ngadi, M.A.; Gungor, V.C. Multi-channel distributed routing scheme for smart grid real-time critical event monitoring applications in the perspective of Industry 4.0. *Int. J. Ad Hoc Ubiquitous Comput.* **2019**, *32*, 236–256. [CrossRef]
10. Dragičević, T.; Lu, X.; Vasquez, J.C.; Guerrero, J.M. DC microgrids—Part II: A review of power architectures, applications, and standardization issues. *IEEE Trans. Power Electron.* **2016**, *31*, 3528–3549. [CrossRef]
11. Yang, Y.; Xu, D.; Ma, T.; Su, X. Adaptive Cooperative Terminal Sliding Mode Control for Distributed Energy Storage Systems. *IEEE Trans. Circuits Syst. I Regul. Pap.* **2021**, *68*, 434–443. [CrossRef]
12. Bevrani, H.; Francois, B.; Ise, T. *Microgrid Dynamics and Control*; John Wiley & Sons, Inc.: Hoboken, NJ, USA, 2017.
13. Wang, C.; Duan, J.; Fan, B.; Yang, Q.; Liu, W. Decentralized High-Performance Control of DC Microgrids. *IEEE Trans. Smart Grid* **2019**, *10*, 3355–3363. [CrossRef]
14. Han, R.; Wang, H.; Jin, Z.; Meng, L.; Guerrero, J.M. Compromised Controller Design for Current Sharing and Voltage Regulation in DC Microgrid. *IEEE Trans. Power Electron.* **2019**, *34*, 8045–8061. [CrossRef]
15. Severino, B.; Strunz, K. Enhancing Transient Stability of DC Microgrid by Enlarging the Region of Attraction Through Nonlinear Polynomial Droop Control. *IEEE Trans. Circuits Syst. I Regul. Pap.* **2019**, *66*, 4388–4401. [CrossRef]
16. Tucci, M.; Meng, L.; Guerrero, J.M.; Ferrari-Trecate, G. Stable current sharing and voltage balancing in DC microgrids: A consensus-based secondary control layer. *Automatica* **2018**, *95*, 1–13. [CrossRef]
17. Han, R.; Tucci, M.; Martinelli, A.; Guerrero, J.M.; Ferrari-Trecate, G. Stability Analysis of Primary Plug-and-Play and Secondary Leader-Based Controllers for DC Microgrid Clusters. *IEEE Trans. Power Syst.* **2019**, *34*, 1780–1800. [CrossRef]
18. Chen, Z.; Yu, X.; Xu, W.; Wen, G. Modeling and Control of Islanded DC Microgrid Clusters with Hierarchical Event-Triggered Consensus Algorithm. *IEEE Trans. Circuits Syst. I Regul. Pap.* **2021**, *68*, 376–386. [CrossRef]
19. Hou, X.; Sun, Y.; Han, H.; Liu, Z.; Yuan, W.; Su, M. A fully decentralized control of grid-connected cascaded inverters. *IEEE Trans. Sustain. Energy* **2019**, *10*, 315–317. [CrossRef]
20. Soliman, H.M.; Bayoumi, E.H.E.; El-Sheikhi, F.A.; Ibrahim, A.M. Ellipsoidal-Set Design of the Decentralized Plug and Play Control for Direct Current Microgrids. *IEEE Access* **2021**, *9*, 96898–96911. [CrossRef]
21. Guo, F.; Xu, Q.; Wen, C.; Wang, L.; Wang, P. Distributed Secondary Control for Power Allocation and Voltage Restoration in Islanded DC Microgrids. *IEEE Trans. Sustain. Energy* **2018**, *9*, 1857–1869. [CrossRef]

22. Amiri, H.; Markadeh, G.R.A.; Dehkordi, N.M.; Blaabjerg, F. Fully decentralized robust backstepping voltage control of photovoltaic systems for DC islanded microgrids based on disturbance observer method. *ISA Trans.* **2020**, *101*, 471–481. [CrossRef] [PubMed]
23. Hu, S.; Yuan, P.; Yue, D.; Dou, C.; Cheng, Z.; Zhang, Y. Attack-Resilient Event-Triggered Controller Design of DC Microgrids Under DoS Attacks. *IEEE Trans. Circuits Syst. I Regul. Pap.* **2020**, *67*, 699–710. [CrossRef]
24. Mehdi, M.; Jamali, S.Z.; Khan, M.O.; Baloch, S.; Kim, C.-H. Robust control of a DC microgrid under parametric uncertainty and disturbances. *Electr. Power Syst. Res.* **2020**, *179*, 106074. [CrossRef]
25. Wan, Y.; Long, C.; Deng, R.; Wen, G.; Yu, X.; Huang, T. Distributed Event-Based Control for Thermostatically Controlled Loads Under Hybrid Cyber Attacks. *IEEE Trans. Cybern.* **2021**, *51*, 5314–5327. [CrossRef] [PubMed]
26. Sahoo, S.; Mishra, S. An Adaptive Event-Triggered Communication-Based Distributed Secondary Control for DC Microgrids. *IEEE Trans. Smart Grid* **2018**, *9*, 6674–6683. [CrossRef]
27. Zhou, J.; Xu, Y.; Sun, H.; Wang, L.; Chow, M.-Y. Distributed Event-Triggered H∞ Consensus Based Current Sharing Control of DC Microgrids Considering Uncertainties. *IEEE Trans. Ind. Inform.* **2020**, *16*, 7413–7425. [CrossRef]
28. Chen, P.; Yu, L.; Zhang, D. Event-Triggered Sliding Mode Control of Power Systems with Communication Delay and Sensor Faults. *IEEE Trans. Circuits Syst. I Regul. Pap.* **2021**, *68*, 797–807. [CrossRef]
29. Sadabadi, M.S.; Shafiee, Q.; Karimi, A. Plug-and-Play Robust Voltage Control of DC Microgrids. *IEEE Trans. Smart Grid* **2018**, *9*, 6886–6896. [CrossRef]
30. Khlebnikov, M.V.; Polyak, B.T.; Kuntsevich, V.M. Optimization of linear systems subject to bounded exogenous disturbances: The invariant ellipsoid technique. *Autom Remote Control* **2011**, *72*, 2227–2275. [CrossRef]
31. Poznyak, A.; Polyakov, A.; Azhmyakov, V. *Attractive Ellipsoids in Robust Control*; Springer International Publishing: Cham, Switzerland, 2014.
32. Awad, H.; Bayoumi, E.H.E.; Soliman, H.M.; De Santis, M. Robust Tracker of Hybrid Microgrids by the Invariant-Ellipsoid Set. *Electronics* **2021**, *10*, 1794. [CrossRef]
33. Soliman, H.M.; El-Sheikhi, F.A.; Bayoumi, E.H.E.; De Santis, M. Ellipsoidal Design of Robust Stabilization for Markov Jump Power Systems under Normal and Contingency Conditions. *Energies* **2022**, *15*, 7249. [CrossRef]
34. Soliman, H.M.; Saleem, A.; Bayoumi, E.H.E.; De Santis, M. Harmonic Distortion Reduction of Transformer-Less Grid-Connected Converters by Ellipsoidal-Based Robust Control. *Energies* **2023**, *16*, 1362. [CrossRef]
35. *Standard IEEE Std 1159-2009 (Revision IEEE Std 1159-1995)*; IEEE Recommended Practice for Monitoring Electric Power Quality. IEEE: Piscataway, NJ, USA, 2009; pp. 1–94. [CrossRef]
36. Ye, Z.; Zhang, D.; Deng, C.; Yan, H.; Feng, G. Finite-time resilient sliding mode control of nonlinear UMV systems subject to DoS attacks. *Automatica* **2023**, *156*, 111170. [CrossRef]

sensors

MDPI

Article

Simplifying Rogowski Coil Modeling: Simulation and Experimental Verification

Alessandro Mingotti *, Christian Betti , Roberto Tinarelli and Lorenzo Peretto

Department of Electrical, Electronic and Information Engineering, Guglielmo Marconi, Alma Mater Studiorum, University of Bologna, Viale del Risorgimento 2, 40136 Bologna, Italy; christian.betti2@unibo.it (C.B.); roberto.tinarelli3@unibo.it (R.T.); lorenzo.peretto@unibo.it (L.P.)
* Correspondence: alessandro.mingotti2@unibo.it

Abstract: The integration of renewable energy sources, electric vehicles, and other electrical assets has introduced complexities in monitoring and controlling power networks. Consequently, numerous grid nodes have been equipped with sensors and complex measurement systems to enhance network observability. Additionally, real-time power network simulators have become crucial tools for predicting and estimating the behavior of electrical quantities at different network components, such as nodes, branches, and assets. In this paper, a new user-friendly model for Rogowski coils is presented and validated. The model's simplicity stems from utilizing information solely from the Rogowski coil datasheet. By establishing the input/output relationship, the output of the Rogowski coil is obtained. The effectiveness and accuracy of the proposed model are tested using both simulations and commercially available Rogowski coils. The results confirm that the model is simple, accurate, and easily implementable in various simulation environments for a wide range of applications and purposes.

Keywords: instrument transformers; accuracy; Rogowski coil; modelling; uncertainty; industrial oriented

Citation: Mingotti, A.; Betti, C.; Tinarelli, R.; Peretto, L. Simplifying Rogowski Coil Modeling: Simulation and Experimental Verification. *Sensors* **2023**, *23*, 8032. https://doi.org/10.3390/s23198032

Academic Editors: Pawel Rozga, Michał Kunicki and Jan Fulneček

Received: 28 August 2023
Revised: 21 September 2023
Accepted: 21 September 2023
Published: 22 September 2023

1. Introduction

In recent years, the network has undergone a significant revolution. On one hand, there have been notable changes in electric assets. For instance, photovoltaic plants are becoming increasingly common on the roofs of industrial, commercial, and private buildings [1,2]. The number of electric vehicles has rocketed in the last few years thanks to commercial maneuvers and country-based incentives [3,4]. Electric scooters have also become a popular means of private transportation, requiring a power supply for charging [5,6]. On the other hand, monitoring techniques and associated software have been evolving. The advancement and widespread adoption of artificial intelligence (AI) and machine learning (ML) techniques have brought about significant changes in power system monitoring techniques [7,8]. The common thread among all these developments is instrumentation. The new assets can be monitored through sensors and distributed measurement systems deployed throughout the power network. These sensors provide measurements for the monitoring system and the learning algorithms used in ML. Therefore, it can be concluded that sensors and instrumentation play a significant role.

Voltage and current measurements are the primary activities conducted at network nodes and branches. However, numerous other quantities are also measured, such as humidity, temperature, power, energy, pressure, etc. [9–11]. All these electrical and environmental parameters are measured to obtain the most accurate health status of the grid. As a result, studies and standards dedicated to sensors and instrumentation are published on a daily basis. Regarding standards, the IEC 61869 series comprises fifteen main documents that pertain to instrument transformers (ITs). Readers can refer to documents IEC 61869-1 and -6 [12,13] or the generic specifications applicable to ITs and low-power instrument transformers (LPITs), respectively. In recent literature, a calibration technique associated with partial discharge measurement using a high-frequency current transformer (HFCT) is described in [14]. ITs with digital output are calibrated in [15]. The authors in [16] characterize a measurement chain that includes LPIT and the phasor measurement unit (PMU). The aim in [17] is to accurately measure distorted signals, while [18] deals with design solutions for ITs, and [19] focuses on failure investigation techniques to be applied to ITs.

In this paper, the IT considered is the Rogowski coil. Due to its features, the Rogowski coil is widely adopted in various applications beyond power systems. Consequently, Rogowski coil modeling has been a research topic for the past few decades, and numerous works can be found in the literature. For example, an electro-thermal model written in VHDL-AMS language and based on finite element analysis (FEA) is described in [20]. In [21], the authors design a Rogowski coil from scratch for three-phase current measurement in electric motors. A highly detailed model is presented in [22], relying on equations that require the complete geometry of the device and the space in which it is installed. A four-layer PCB-based Rogowski coil is designed in [23]. A simple model, although based on several preliminary measurements, is detailed in [24]. The same authors introduced a new testing signal, known as the sinc response, to be measured by the Rogowski coil to obtain its input-output response [25]. The high-frequency behavior of the Rogowski coil is studied in [26] using the classical lumped model, while [27] explores another high-frequency modeling approach based on a black-box conceptualization of the Rogowski derived from preliminary measurements on the device.

From the literature review, it becomes apparent that there is a lack of very simple models that do not require extensive preliminary measurements. Furthermore, obtaining information on the internal structure and design of the device is not always feasible. This makes the modeling even more complicated and not generalizable for every commercial Rogowski coil. To address this issue, this paper introduces a new and straightforward method to obtain the frequency model of a commercial Rogowski coil. In practice, it is common to purchase a device for various applications rather than designing a custom Rogowski coil from scratch. The proposed model is based on a few inputs obtained from the device's datasheets, and a set of equations is developed to derive a transfer function dependent on the cross-sectional shape. The lack of knowledge about commercial devices' internal structure is an obstacle to many existing models. Subsequently, the obtained model is implemented in the Matlab 2023a environment for preliminary efficacy testing. Finally, the model is experimentally validated through measurements performed on commercial Rogowski coils. The results clearly demonstrate the validity and accuracy of the frequency model, confirming its potential for implementation in simulation environments such as real-time power network simulators.

The remainder of this paper is structured as follows: Section 2 includes the theoretical concepts and motivation behind the work. Section 3 is dedicated entirely to the model description and contextualization. Model validation is presented in Section 4. Finally, Section 5 concludes this paper and provides suggestions for future research.

2. Motivation and Background

2.1. The Rogowski Coil

The Rogowski coil (RC) is an LPIT for the measurement of AC and, with some expedients, also DC current. Manufacturers can refer to the standard IEC 61869-10 [28] for designing details, testing procedures, and accuracy evaluation. RC is widely adopted for power system applications due to its features. They are small, lightweight, and free from iron cores, and often they can be opened to facilitate their installation. Considering other current measurement solutions, e.g., inductive current transformers (CT), the RC features become significant.

The RC working principle is quite simple, and it can be described looking at Figure 1.

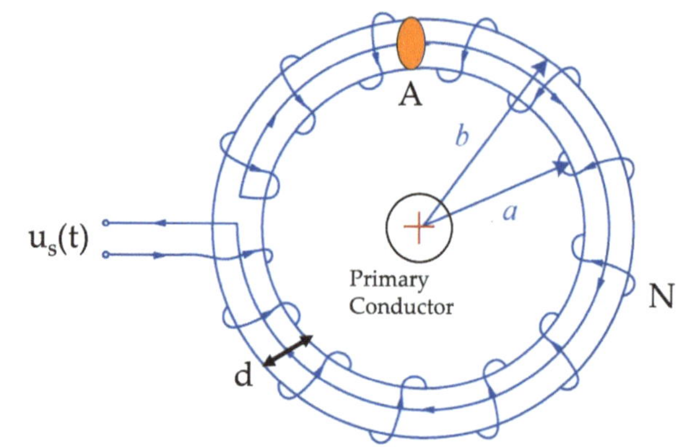

Figure 1. Picture of the Rogowski coil and its components.

A copper wire is wound around an iron-free insulating support. This prevents any saturation behavior, typical of iron-core-based devices. The conductor, whose current I_P must be measured, is inserted inside the RC. Its output, a voltage, $u_S(t)$, is proportional to the derivative of I_P:

$$u_S(t) = -M\frac{dI_P}{dt} \quad (1)$$

where M is the mutual inductance between the primary and secondary windings. The input-output relationship of the RC raises some comments. On the one hand, being the output voltage, it becomes easier to connect the RC to common analog-to-digital converters (ADC) without any other element in the measurement chain. On the other hand, u_S is proportional to the derivative of the current and not to the current itself. Therefore, to obtain the correct value of I_P, an analog or digital integration step is required. As for this paper, the integration system is not considered. The aim is to model the RC behavior independently of the integration solution adopted (it might introduce many aspects to be considered to obtain the input-output relationship of the RC).

2.2. Motivation

The need for RC models is supported by their massive adoption in several applications in different fields. As previously mentioned, many models are already available; however, their implementation is typically not straightforward. Therefore, the main goal of this paper is to provide an easy-to-use modeling procedure to obtain the input-output relationship of whatever commercial (or not) RC. The modeling procedure does not involve any preliminary electrical measurements on the RC but only external geometrical ones. Of course, the availability of preliminary information is an added value for the modeling procedure.

The benefits of a simplified modeling procedure are several. For example, an a priori estimate of the RC response would be possible without performing a time-consuming laboratory measurement. Second, the use of digital simulators is spreading; hence, having an accurate RC model would allow for the integration of realistic current measurements inside the simulations. Finally, the optimum exploitation of a model is the creation of a RC digital twin, which would help from the design to the implementation of the RC inside each application.

3. The Modelling Procedure

The proposed modeling procedure starts with the geometrical measurements on the RC and ends with the transfer function (TF) that correlates the frequency relation between the output voltage and the input current. The procedure is summarized in the flowchart depicted in Figure 2.

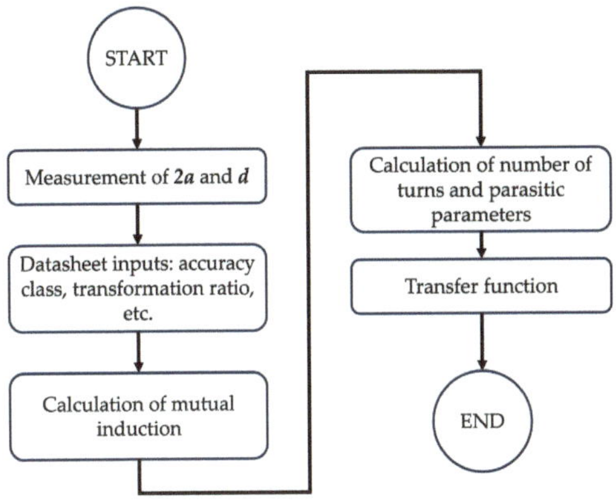

Figure 2. Flowchart of the modeling procedure.

The first step is to measure the inner diameter 2*a* and the diameter of the cross-section *d*. The frequency modeling procedure is developed for three cross-section geometries: circular, square/rectangular, and oval. In the latter two geometries, an extra measurement is needed, as explained in Figure 3. As for the oval case, commercial devices do not present ideal oval shapes. Therefore, the geometry is often schematized, as shown in Figure 3.

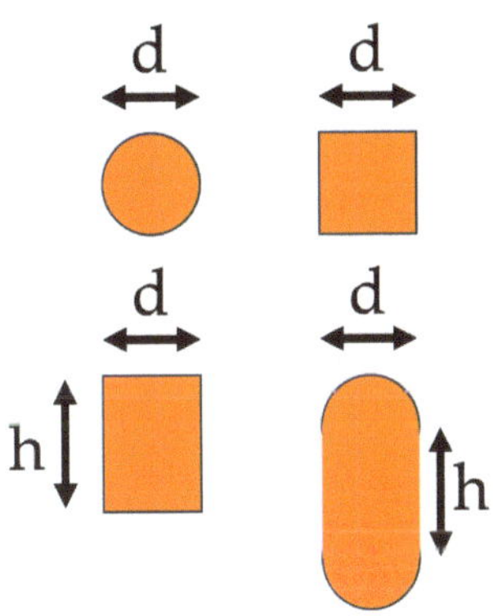

Figure 3. Cross-section geometries.

The second step consists of extracting some information from the datasheet. The two key parameters needed for the modeling are the accuracy class and the transformation ratio (TR). Other typical parameters, given in the RC datasheet, are listed in Table 1. Note that the TF also exploits the information about the rated burden. However, if not provided by the manufacturer, the standard value of 2 MΩ given in [13] can be used.

Table 1. Typical parameters available in the RC datasheet.

Accuracy Class	Linearity Error	Resistance of the Windings	Transformation Ratio
Max Current	Bandwidth	Temperature Class	Rated Burden

The third step allows us to obtain the mutual induction, *M* starting from (1). Substituting the derivative over time with the angular speed $\omega = 2\pi f$:

$$M = \frac{TR_{50}}{\omega} \tag{2}$$

where f is the generic frequency and TR_{50} is the rated transformation ratio (at 50 Hz). Once M is found, it is possible to obtain the TR at each frequency of interest. For this assumption, the parasitic parameters will be neglected.

The fourth step regards geometry. Using the quantities obtained in the first step, the perimeter and the cross-section area of the RC can be calculated.

At this point, using consolidated expressions of the mutual inductance [29,30], the number of turns of the RC can be obtained from the reversed formula. Note that, from a practical perspective, the lack of knowledge of the number of turns is one of the limiting factors of RC modeling. Hence:

$$N_{rec} = \frac{2\pi M}{d\mu log\left(\frac{2a+2d}{2a}\right)} \tag{3}$$

$$N_{cir} = \frac{2M}{\mu\left(2a + d - 2\sqrt{(a+b)a}\right)} \tag{4}$$

$$N_{ova} = \frac{2\pi M}{\mu\left(2a + d - 2\sqrt{(a+b)a} + \frac{h}{\pi}\log\left(\frac{a+b}{a}\right)\right)} \tag{5}$$

where μ is the overall permeability, and N_{rec}, N_{cir}, and N_{ova} are the number of turns of the rectangular, circular, and oval cross-section, respectively. They can be used to obtain, in addition to the geometrical parameters, the length l_w of the single coil, section A_w, and radius r_w of the wire used to wound the RC.

The last step is the calculation of the parasitic parameters of the RC:

$$C_i = \frac{2\varepsilon\pi^2(2a+d)}{\log\left(\frac{\sqrt{\frac{A_i}{\pi}}}{r_w}\right)} \tag{6}$$

$$L_{rec} = \frac{\mu d log\left(\frac{a+d}{a}\right)N_{rec}^2}{2\pi} \tag{7}$$

$$L_{cir} = \frac{\mu N_{cir}^2\left(2a + d - 2\sqrt{(a+d)a}\right)}{2} \tag{8}$$

$$L_{ova} = \frac{\mu N_{ova}^2\left(2a + d - 2\sqrt{(a+d)a} + \frac{h}{\pi}\log\left(\frac{a+d}{a}\right)\right)}{2\pi} \tag{9}$$

where ε is the overall permittivity (to be fixed depending on the used material), C and L are the parasitic capacitance and inductance, respectively. The suffix i is used to switch among the cross-section parameters (A_i). As for the winding resistance, R_w, it can be easily obtained with the second Ohm's law, measured, or extracted from the datasheet. Of course, there will be datasheets with extra information that can be used instead of some parameter expressions. However, the proposed method has been generalized to avoid any exceptional cases of missing (extra) inputs. A small note on permeability and permittivity is necessary. The general notation has been used to highlight that, depending on the RC adopted, the user may need to insert or not insert the relative permeability/permittivity. Further tests on these specific parameters demonstrated an almost negligible effect of the permittivity/permeability value on the accuracy parameters.

With the results from (2)–(9), the RC transfer function in the "s" domain can be obtained as:

$$TF = \frac{ZMs}{ZCLs^2 + s(L + R_wCZ) + R_w + Z} \tag{10}$$

where Z is the rated burden, or better, the impedance of the measuring device connected in cascade to the RC. The *TF* in (10) must be customized with the parameters associated with the studied cross-section. To better contextualize the *TF* in (10), a schematic with the equivalent circuit is depicted in Figure 4. All the symbols in the picture were previously described.

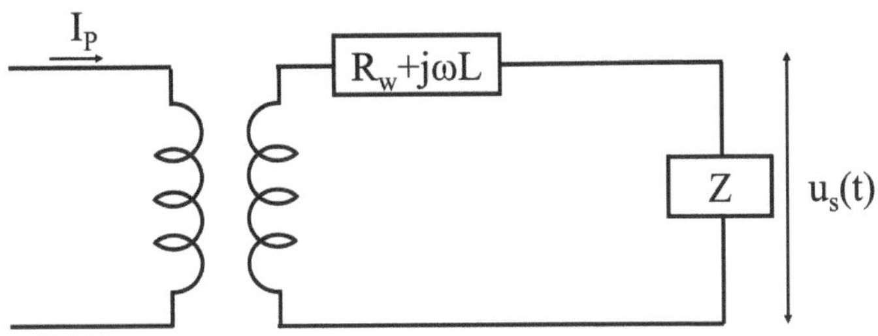

Figure 4. Schematic of the equivalent circuit of the RC.

4. Validation

In this section, two types of validations are presented. The first validation is obtained by simulation, and the second by experimental measurements on commercial RCs.

4.1. Validation by Simulation

To verify the performance of the proposed method, two sets of tests are performed. First, a frequency sweep test is simulated to assess the accuracy vs. frequency. Second, the MonteCarlo method (MCM) is applied to quantify the effect of the sources of uncertainty on the accuracy of the method.

4.1.1. Frequency Sweep

The simulation of the frequency sweep was run in the Matlab 2023 environment. A sinusoidal signal with an amplitude of 1000 A and a 0° initial phase was generated for all the frequencies from 50 Hz to 2500 Hz at 50 Hz steps. The frequency range was selected to cover the power-quality frequency range. As for the acquisition details, the input currents were simulated using a 200 ms window and a 50 kSa/s sampling frequency. Afterwards, each signal was multiplied by the *TF* in (10) to obtain the RC response. By means of the discrete Fourier transform (DFT), the goodness of the RC response was evaluated with the typical parameters: ratio error ε and phase displacement $\Delta\varphi$:

$$\varepsilon = \frac{TR_f I_{Sf} - I_{Pf}}{I_{Pf}} 100 \tag{11}$$

$$\Delta\varphi = \hat{I_{Sf}} - \hat{I_{Pf}} \tag{12}$$

where TR_f, I_{Sf}, and I_{Pf} are the transformation ratio, the rms of the secondary current, and the rms of the primary current, respectively, evaluated at the generic frequency f. Analogously, $\hat{I_{Sf}}$ and $\hat{I_{Pf}}$ are the phase angles of the secondary and primary currents, respectively.

The results of this first set of tests are listed in Table 2 and depicted in Figure 5. As for the RC simulated, it is a rectangular cross-section, 0.5 accuracy class device.

From the results, it can be concluded that the estimation of the RC output is accurate in all the PQ frequency ranges. In terms of absolute value, the results are aligned with the accuracy class of the RC simulated. For the limits of ε and $\Delta\varphi$, the reader can refer to [13], which provides the maximum values allowed for each harmonic up to the 13th.

Table 2. ε and $\Delta\varphi$ for each test signal.

Frequency (Hz)	ε (%)	$\Delta\varphi$ (mrad)
50	−0.01	3.2
100	−0.01	6.3
150	−0.01	9.5
200	−0.01	12.6
250	−0.01	15.8
300	−0.02	18.9
350	−0.02	22.1
400	−0.02	25.2
450	−0.02	28.4
500	−0.03	31.5
550	−0.03	34.7
600	−0.03	37.8
650	−0.04	41.0
700	−0.04	44.1
750	−0.05	47.3
800	−0.05	50.4
850	−0.06	53.6
900	−0.06	56.7
950	−0.07	59.9
1000	−0.08	63.0
1050	−0.08	66.2
1100	−0.09	69.3
1150	−0.1	72.5
1200	−0.11	75.6
1250	−0.11	78.8
1300	−0.12	81.9
1350	−0.13	85.1
1400	−0.14	88.2
1450	−0.15	91.4
1500	−0.16	94.5
1550	−0.17	97.7
1600	−0.18	100.9
1650	−0.19	104.0
1700	−0.2	107.2
1750	−0.21	110.3
1800	−0.23	113.5
1850	−0.24	116.6
1900	−0.25	119.8
1950	−0.26	122.9
2000	−0.28	126.1
2050	−0.29	129.2
2100	−0.3	132.4
2150	−0.32	135.5
2200	−0.33	138.7
2250	−0.35	141.8
2300	−0.36	145.0
2350	−0.38	148.1
2400	−0.39	151.3
2450	−0.41	154.4
2500	−0.43	157.6

4.1.2. Sources of Uncertainty

To further validate a method, it is good practice to locate and assess the sources of uncertainty. For the proposed approach, the geometrical measures are those affected by the measurement uncertainty. It is worth mentioning that many influence quantities and many sources of uncertainty might affect the performance of RCs during normal operations. However, the uncertainty analysis run in this paper is focused on the sources that directly affect the proposed procedure. Therefore, the quantities d, h (Figure 3), and $2a$ are those corrupted by uncertainty in the MCM implementation. The MCM was implemented with ten thousand iterations, assuming a uniform distribution for the probability density functions associated with the quantities. Considering realistic measurement devices used to measure lengths, the limits of the uniform distribution were fixed at 1%, 10%, and 20% of the measured quantity. To consider the worst case, the measurement of the RC windings' resistance was also considered. Therefore, the same percentages and distributions apply to the resistance value R_w. The results of the MCM implementation are listed in Table 3. For the sake of brevity, the case of the rectangular cross-section is considered.

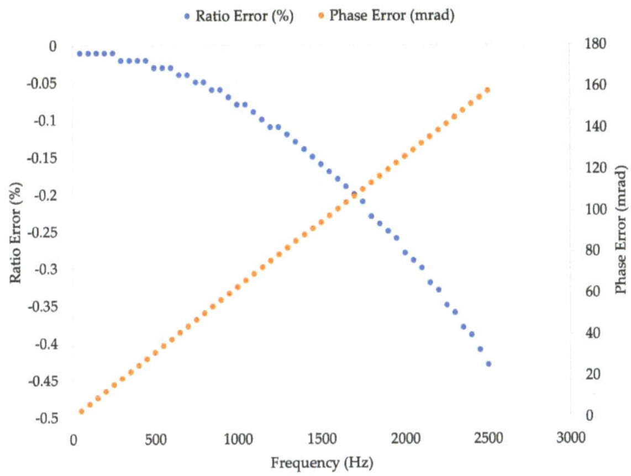

Figure 5. ε and $\Delta\varphi$ for each test signal described in Table 2.

Table 3. Results of the uncertainty analysis.

Uncertainty Considered (%)	Quantity	μ	$l_{95\%}$	$u_{95\%}$
1	N (-)	1340	1309	1370
10	N (-)	1350	1054	1665
20	N (-)	1402	837	2092
1	C (F)	3.8×10^{-12}	3.8×10^{-12}	3.9×10^{-12}
10	C (F)	3.9×10^{-12}	3.4×10^{-12}	4.3×10^{-12}
20	C (F)	3.9×10^{-12}	3.0×10^{-12}	4.6×10^{-12}
1	L (H)	4.3×10^{-4}	4.2×10^{-4}	4.4×10^{-4}
10	L (H)	4.3×10^{-4}	3.3×10^{-4}	5.3×10^{-4}
20	L (H)	4.5×10^{-4}	2.7×10^{-4}	6.6×10^{-4}

For each quantity affected by uncertainty and for each percentage of uncertainty, the table contains the mean value μ, the lower limit $l_{95\%}$ and the upper limit $u_{95\%}$ of the 95% confidence interval. As expected, the spread of the results increases with the increase in uncertainty associated with the length measurements. However, the parasitic parameters are slightly affected by such a contribution. The most affected parameter is the number of turns, which is already well estimated by the method. For the sake of clarity, in Figure 6, the pdf of the parasitic capacitance C obtained running 10,000 Monte Carlo trials is plotted. Considering that C is obtained from the combination of more than one random variable affected by uncertainty, the resulting pdf is prone to a trapezoidal one.

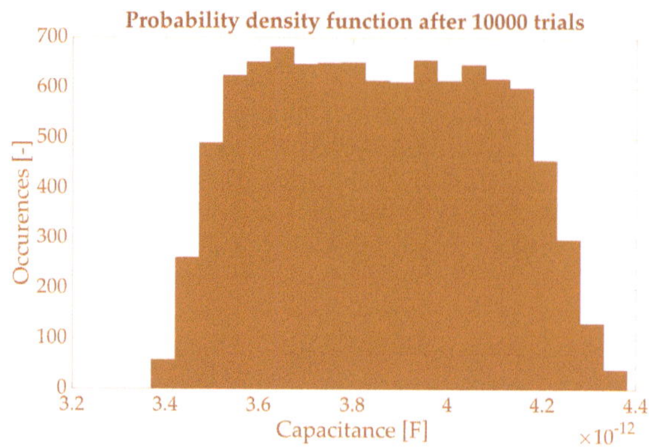

Figure 6. Probability density function obtained for C after 100,000 Monte Carlo trials.

4.2. Validation by Measurements

To assess the usefulness and applicability of the proposed approach, it is fundamental to perform validation with commercial devices. To this purpose, three RCs, namely R1, R2, and R3, were tested. Table 4 lists the characteristics of the RCs.

Table 4. Characteristics of the commercial RCs under test.

Feature	R1	R2	R3
TR_{50} (mV/kA)	100	100	100
d (mm)	8	8	12
a (mm)	50	57	21.5
R_w (Ω)	256	381	22
Cross-Section	Circular	Circular	Oval
Accuracy Class	0.5	1	1
Bandwidth	1 Hz to 100 kHz	NA	20 Hz to 5 kHz
Operating Temperature	−30 °C to 80 °C	−20 °C to 85 °C	−20 °C to 70 °C
Rated Current (A)	1000	10,000	1000

The RC were tested with the measurement setup depicted in Figure 7. It includes a Fluke calibrator 6105A and its transconductance Fluke 52120A. They were used together to generate the desired current a reference shunt to measure the input current injected through the RCs. The shunt resistance is, after an accurate characterization, 1.02129 mΩ ± 0.00008 mΩ, up to 2500 Hz. A data acquisition board, NI 9238, whose characteristics are given in Table 5, it was used to collect the three voltage signals from the RCs under test and the reference signal coming from the reference shunt. In Figure 7, a color code is used to distinguish between the current and the voltage signals.

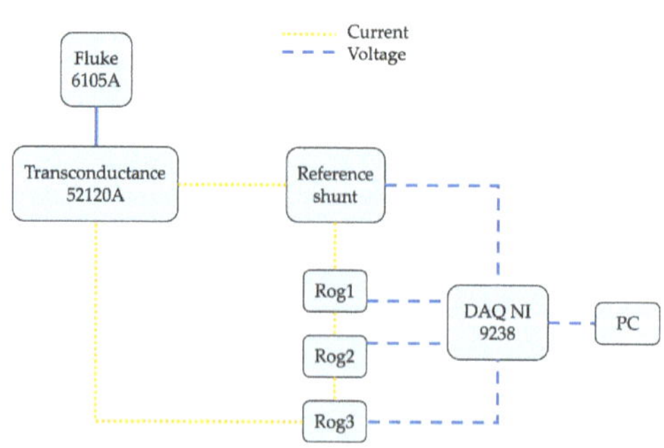

Figure 7. The measurement setup used for the experimental measurements on the commercial RCs.

Table 5. Main characteristics of the data acquisition board NI9238.

Converter	24-bit	**Voltage Range**	±500 mV
Sampling Frequency	50 kSa/s/Ch	**Input Impedance**	>1 GΩ
Simultaneous Channels	Yes	**Gain Error**	±0.07%
Offset Error	±0.005%	**Input Noise**	3.9 μV

The tests performed with the setup in Figure 7 are here described. The signals described in Table 2 were generated with the Fluke plus transconductance and injected through the reference shunt and the RCs under test. All tests were performed using a primary current of 100 A. The load impedance is given in Table 5, and it is greater than 1 GΩ. This preliminary measurement allows the collection of the actual RC response. In fact, the second part of the test consists of the combination of the injected signal (acquired from the shunt) and the TFs obtained from the modeling approach. The result is an estimation of the RC behavior at each frequency. To validate the estimation process, hence the proposed modeling approach, the phase error (PE) and the percentage voltage difference (VD) indicators are used. The PE is the difference in mrad between the phase displacement obtained in simulation and the one obtained during the experimental measurements (it is not the difference between absolute values; it would be meaningless). The VD, instead, is merely the percentage difference between the voltage measured by the RC and the one estimated by the method (using the rated current as the base of the ratio). The results, for the sake of brevity, are listed in Table 6 for a comprehensive selection of frequencies.

Table 6. Results from the comparison of the actual and estimated RC behaviors.

Frequency (Hz)	R1		R2		R3	
	PE (mrad)	VD (%)	PE (mrad)	VD (%)	PE (mrad)	VD (%)
50	2.92	0.4	−2.80	0.5	−4.41	1.8
250	14.36	1.2	−12.83	1.0	−15.88	4.2
550	31.39	2.7	−26.91	1.5	−33.66	8.9
850	48.34	4.6	−40.71	1.7	−51.73	13.5
1250	70.93	5.3	−59.10	0.8	−75.91	13.7
2500	141.77	15.8	116.92	−5.1	−151.31	22.8

According to the results, it is possible to appreciate how accurately the model can estimate RC behavior. As expected, the accuracy drops when the frequency increases. This is due to the higher uncertainty of the generation system and the lower number of samples per period at those frequencies. However, for all RCs and at all frequencies, the obtained errors are far below the limits given in [13]. The only exception is the VD at 2500 Hz. The limit given in [13] is 20%. However, at this stage, and considering the amount of approximation conducted in the simplified procedure, the results can be considered promising. It is worth mentioning that, even if the results are all above the uncertainty limits (except one), the goodness of the results is increased by the final application in which they are going to be used. For example, applications with limited target uncertainty will benefit the most from the proposed method.

5. Discussion

The obtained results trigger discussion on the proposed method. As a matter of fact, the simulations performed and the experimental measurements provided encouraging results. They were obtained by exploiting almost any input but those included in the datasheet. The direct consequence is obvious. The model would provide far better results if one or more measurements could be performed on the RCs. Furthermore, the modeling approach can be improved if extra information is added to it. For example, imagine a RC coil being characterized vs. temperature. Its results can be included in the model, which will differentiate its output depending on the working temperature. The same consideration can be extended to any other influence quantity that may affect the RC operation. Therefore, the chances of exploitation of the modeling approach are several, such as the simple laboratory, the company that develops power system simulators, the metrological institute, etc.

6. Conclusions

This paper aims to provide a simplified modeling approach to be applied to Rogowski coils. It is designed to be applied at all levels, with the flexibility to be improved depending on the available information. The approach is then tested with both simulations and experimental measurements. The results clearly confirm the validity of the approach and pave the way for further studies needed for its improvement.

Author Contributions: Conceptualization, A.M.; methodology, R.T.; software, C.B.; validation, L.P.; formal analysis, A.M.; investigation, C.B.; resources, L.P.; data curation, C.B.; writing—original draft preparation, A.M.; writing—review and editing, R.T. All authors have read and agreed to the published version of the manuscript.

Funding: This research received no external funding.

Conflicts of Interest: The authors declare no conflict of interest.

References

1. Ahmed, Y.A.; Rashid, A.; Khurshid, M.M. Investigating the determinants of the adoption of solar photovoltaic Systems—Citizen's perspectives of two developing countries. *Sustainability* **2022**, *14*, 1764. [CrossRef]
2. Linhart, M.; Rodin, V.; Moser, S.; Kollmann, A. Citizen participation to finance pv power plants focused on self-consumption on company roofs-findings from an austrian case study. *Energies* **2021**, *14*, 738. [CrossRef]
3. Breschi, V.; Ravazzi, C.; Strada, S.; Dabbene, F.; Tanelli, M. Fostering the mass adoption of electric vehicles: A network-based approach. *IEEE Trans. Control Netw. Syst.* **2022**, *9*, 1666–1678. [CrossRef]
4. Lv, Z.; Qiao, L.; Cai, K.; Wang, Q. Big data analysis technology for electric vehicle networks in smart cities. *IEEE Trans. Intell. Transp. Syst.* **2021**, *22*, 1807–1816. [CrossRef]
5. Hieu, L.T.; Lim, O.T. Effects of the structure and operating parameters on the performance of an electric scooter. *Sustainability* **2023**, *15*, 8976. [CrossRef]
6. Kubik, A. The energy consumption of electric scooters used in the polish shared mobility market. *Energies* **2022**, *15*, 8193. [CrossRef]
7. Razmi, P.; Ghaemi Asl, M. Machine learning approaches in a real power system and power markets. In *Application of Machine Learning and Deep Learning Methods to Power System Problems*; Power Systems; Springer: Cham, Switzerland, 2021. [CrossRef]
8. Zhu, Y.; Singh, C. End-to-end topology-aware machine learning for power system reliability assessment. In Proceedings of the 2022 17th International Conference on Probabilistic Methods Applied to Power Systems, PMAPS 2022, Manchester, UK, 12–15 June 2022. [CrossRef]
9. Chen, X.; Xu, J.; Fang, Y.; Li, W.; Ding, Y.; Wan, Z.; Tu, Z. Temperature and humidity management of PEM fuel cell power system using multi input and multi-output fuzzy method. *Appl. Therm. Eng.* **2022**, *203*, 117865. [CrossRef]
10. Wu, W.; Yue, S.; Zhou, X.; Guo, M.; Wang, J.; Ren, L.; Yuan, B. Observational study on the impact of large-scale photovoltaic development in deserts on local air temperature and humidity. *Sustainability* **2020**, *12*, 3403. [CrossRef]
11. Peretto, L.; Tinarelli, R.; Ghaderi, A.; Mingotti, A.; Mazzanti, G.; Valtorta, G.; Danesi, S. Monitoring cable current and laying environment parameters for assessing the aging rate of MV cable joint insulation. In Proceedings of the Annual Report—Conference on Electrical Insulation and Dielectric Phenomena, CEIDP, Shenzhen, China, 20–23 October 2018; pp. 390–393. [CrossRef]
12. *IEC 61869-1:2010*; Instrument Transformers—Part 1: General Requirements. International Standardization Organization: Geneva, Switzerland, 2010.
13. *IEC 61869-6*; Instrument Transformers—Part 6: Additional General Requirements for Low-Power Instrument Transformers. International Standardization Organization: Geneva, Switzerland, 2016.
14. Besharatifard, H.; Hasanzadeh, S.; Muyeen, S.M.; Kamwa, I. Evaluation of a calibration technique in measuring partial discharges inside mineral oils with a high-frequency current transformer (HFCT) sensor: A case study. *IET Gener. Transm. Distrib.* **2023**, *17*, 706–715. [CrossRef]
15. Chen, Y.; Crotti, G.; Dubowik, A.; Letizia, P.S.; Mohns, E.; Luiso, M.; Bruna, J. Novel calibration systems for the dynamic and steady-state testing of digital instrument transformers. In Proceedings of the AMPS 2021—2021 11th IEEE International Workshop on Applied Measurements for Power Systems, Virtual, 29 September–1 October 2021. [CrossRef]
16. Mingotti, A.; Peretto, L.; Tinarelli, R.; Angioni, A.; Monti, A.; Ponci, F. A simple calibration procedure for an LPIT plus PMU system under off-nominal conditions. *Energies* **2019**, *12*, 4645. [CrossRef]

17. Kaczmarek, M.; Stano, E. Challenges of accurate measurement of distorted current and voltage in the power grid by conventional instrument transformers. *Energies* **2023**, *16*, 2648. [CrossRef]
18. Lesniewska, E. Modern methods of construction problem solving in designing various types of instrument transformers. *Energies* **2022**, *15*, 8199. [CrossRef]
19. Tee, S.; Liu, Q.; Wang, Z.; Hafid, F.; Tournet, P. Failure investigation and asset management of combined measuring instrument transformers. *High Volt.* **2021**, *6*, 61–70. [CrossRef]
20. Estupiñán JS, R.; Vachoux, A.; Pascal, J. Electro-thermal modeling of a rogowski coil sensor system. In Proceedings of the 2015 International Symposium on VLSI Design, Automation and Test, VLSI-DAT, Hsinchu, Taiwan, 27–29 April 2015. [CrossRef]
21. Ibrahim, M.E.; Abd-Elhady, A.M. Design and modeling of a two-winding rogowski coil sensor for measuring three-phase currents of a motor fed through a three-core cable. *IEEE Sens. J.* **2021**, *21*, 8289–8296. [CrossRef]
22. Liu, X.; Huang, H.; Jiao, C. Modeling and analyzing the mutual inductance of rogowski coils of arbitrary skeleton. *Sensors* **2019**, *19*, 3397. [CrossRef] [PubMed]
23. Ming, L.; Xin, Z.; Liu, W.; Loh, P.C. Structure and modelling of four-layer screenreturned PCB rogowski coil with very few turns for high-bandwidth SiC current measurement. *IET Power Electron.* **2020**, *13*, 765–775. [CrossRef]
24. Mingotti, A.; Peretto, L.; Tinarelli, R. A simple modelling procedure of rogowski coil for power systems applications. In Proceedings of the AMPS 2019—2019 10th IEEE International Workshop on Applied Measurements for Power Systems, Aachen, Germany, 25–27 September 2019. [CrossRef]
25. Mingotti, A.; Peretto, L.; Tinarelli, R. A smart frequency domain-based modeling procedure of rogowski coil for power systems applications. *IEEE Trans. Instrum. Meas.* **2020**, *69*, 6748–6755. [CrossRef]
26. Orosz, T.; Tamus, Z.A.; Vajda, I. Modeling the high frequency behavior of the rogowski-coil passive L/r integrator current transducer with analytical and finite element method. In Proceedings of the Universities Power Engineering Conference, Cluj-Napoca, Romania, 2–5 September 2014. [CrossRef]
27. Sabiha, N.A.; Alkhammash, H.I. Modeling rogowski coils for monitoring surge arrester discharge currents. *Comput. Syst. Sci. Eng.* **2021**, *42*, 439–449. [CrossRef]
28. *IEC 61869-10*; Instrument Transformers—Part 10: Additional Requirements for Low-Power Passive Current Transformers. International Standardization Organization: Geneva, Switzerland, 2017.
29. Styblìkovà, R.; Draxler, K. Design and calibration of Rogowski coils. In Proceedings of the Power and Energy Workshop, Noordwijk, The Netherlands, 22–23 March 2011.
30. Moreno, M.V.R.; Robles, G.; Albarracín, R.; Rey, J.A.; Tarifa, J.M.M. Study on the self-integration of a Rogowski coil used in the measurement of partial discharges pulses. *Electr. Eng.* **2017**, *99*, 817–826. [CrossRef]

sensors

MDPI

Article

Smart Grid Outlier Detection Based on the Minorization–Maximization Algorithm

Lina Qiao [1,2], Wengen Gao [1,2,*], Yunfei Li [1,2], Xinxin Guo [1,2], Pengfei Hu [1,2] and Feng Hua [1,2]

1 College of Electrical Engineering, Anhui Polytechnic University, Wuhu 241000, China; qln1017@163.com (L.Q.); lyf@mail.ahpu.edu.cn (Y.L.); guoxinxin@ahpu.edu.cn (X.G.); h382024188@163.com (P.H.); huafeng980525@163.com (F.H.)
2 Key Laboratory of Advanced Perception and Intelligent Control of High-End Equipment, Chinese Ministry of Education, Wuhu 241000, China
* Correspondence: ahpuchina@ahpu.edu.cn

Abstract: Outliers can be generated in the power system due to aging system equipment, faulty sensors, incorrect line connections, etc. The existence of these outliers will pose a threat to the safe operation of the power system, reduce the quality of the data, affect the completeness and accuracy of the data, and thus affect the monitoring analysis and control of the power system. Therefore, timely identification and treatment of outliers are essential to ensure stable and reliable operation of the power system. In this paper, we consider the problem of detecting and localizing outliers in power systems. The paper proposes a Minorization–Maximization (MM) algorithm for outlier detection and localization and an estimation of unknown parameters of the Gaussian mixture model (GMM). To verify the performance of the method, we conduct simulation experiments by simulating different test scenarios in the IEEE 14-bus system. Numerical examples show that in the presence of outliers, the MM algorithm can detect outliers better than the traditional algorithm and can accurately locate outliers with a probability of more than 95%. Therefore, the algorithm provides an effective method for the handling of outliers in the power system, which helps to improve the monitoring analyzing and controlling ability of the power system and to ensure the stable and reliable operation of the power system.

Keywords: outlier detection; the Minorization–Maximization algorithm; localization; power system

Citation: Qiao, L.; Gao, W.; Li, Y.; Guo, X.; Hu, P.; Hua, F. Smart Grid Outlier Detection Based on the Minorization–Maximization Algorithm. *Sensors* **2023**, *23*, 8053. https://doi.org/10.3390/s23198053

Academic Editors: Pawel Rozga, Michał Kunicki and Jan Fulneček

Received: 5 August 2023
Revised: 17 September 2023
Accepted: 18 September 2023
Published: 24 September 2023

1. Introduction

With the rapid development of the power industry around the world, the scale of the power system has gradually expanded, and its operation mode and network structure have become more complex. The combination of power grids and modern information forms smart grids, which pose greater challenges to the safety and reliability of power system operations. In order to improve the operational efficiency and reliability of the power grid, the dispatching system needs to collect complete and reliable real-time data for processing to facilitate online analysis and decision control of advanced application software.

Power system state estimation is one of the core functions of the Energy Management System (EMS) in the power system dispatching center [1–3]. Its function refers to the control center to collect various measurement data through sensors and to estimate the current operating state of the power system according to the measurement data. The safe and economical operation of modern power grids depends on the EMS. The many functions of the Energy Management System can be divided into two parts: online application for real-time change analysis of the power grid and offline application for typical power flow section analysis. Power system state estimation is the core of power system online monitoring, analysis, and control functions and plays an important role in the intelligent analysis of power grid dispatching. State estimation is the basis of most advanced software for online applications, and the accuracy of the state estimation results is closely related to the accuracy of the subsequent analysis calculation results. Due to the continuous improvement in the automation of the power system, the quality requirements for real-time measurement data in state estimation are becoming higher and higher. In addition to the measurement noise, there may be outliers in the obtained measurement data. Outliers [4–6] can be caused by a variety of factors. The following are some common causes of outliers:

(1) Sensors can malfunction or fail, resulting in inaccurate or incomplete data collection, resulting in outliers.
(2) Human factors, such as incorrect data entry, operational errors, etc., may cause data anomalies.
(3) Natural disasters and emergencies, such as storms, earthquakes, fires, explosions, etc., may cause damage to electric power facilities and thus generate outliers.
(4) Factors such as power equipment failure, damage, data transmission, and processing errors can lead to outliers.

The existence of outliers will cause the state estimation results to deviate significantly from the actual operation of the system, and the performance of the estimator will be seriously degraded, so the work of outlier detection becomes particularly important.

To solve the problem of outliers in measurement data, the traditional detection methods include weighted least squares (WLS) [7–9], residual search method [10,11], and non-quadratic criterion method [12,13]. These methods are further elaborated on below. The basic idea of the WLS is to adjust the weights of measurements by assigning larger weights to relatively reliable and accurate measurements and smaller weights to measurements that are relatively unreliable and affected by outliers. By adjusting the weights, the WLS algorithm can reduce the effect of outliers on the fitted results and improve the accuracy of the parameter estimates. However, it is important to note that the WLS algorithm has some limitations when dealing with outliers. When the number of outliers is high or the difference between the outliers and normal values is large, the WLS algorithm may not be able to eliminate the effect of the outliers. The residual search method uses weighted residuals and standard residuals to sort the measurement data. After eliminating the data with large residuals, the state estimation is re-estimated to achieve the purpose of optimal estimation. However, the disadvantage of this method is that the calculation amount is large, and it is easy to have residual pollution and residual submergence, resulting in false detection or missed detection. The strategy adopted by the non-quadratic criterion method is to successively reduce the weight of suspicious data, rather than eliminating suspicious data one by one. Re-estimation is avoided to reduce the amount of calculation. However, this may lead to difficult consequences of convergence, and there is no guarantee that the final estimate will be optimal. In summary, traditional outlier detection algorithms have certain limitations in addressing outlier problems. Therefore, new methods have been proposed to solve the problem of the presence of outliers.

In this paper, the Minorization–Maximization (MM) algorithm [14–20] is used to detect outliers. Compared with traditional methods for detecting outliers, the MM algorithm can detect outliers more accurately, especially for complex data distributions, and can obtain better performance. The MM algorithm is an optimization algorithm whose basic idea is to optimize the original function to be optimized indirectly by optimizing the alternative function by finding an easier function to be optimized. In each iteration, the algorithm optimizes the original function by transforming the original problem into a more tractable problem through a series of derivations and transformations. The algorithm first appeared in the field of online search in the 1970s [21], when Ortega and Rheinboldt first mentioned the MM principle [22]. Later, de Leeuw proposed the first MM algorithm in a multidimensional scale analysis, and Hunter and Lange named it the MM algorithm [23], which has since been widely used in statistics. MM algorithms are an important tool in optimization problems because of their conceptual simplicity, ease of implementation, and numerical stability. Moreover, MM algorithms are provably convergent optimization algorithms that are guaranteed to find either a globally optimal solution or a locally optimal solution under certain conditions. Overall, the MM algorithm is a very important optimization algorithm with a wide range of application scenarios and excellent properties that can effectively solve many practical problems.

This paper is divided into six main sections. First, in Section 2, the normal measurement model and the measurement model containing outliers are described. Then, Section 3 introduces the basic principles of the MM algorithm, including the iterative process of the MM algorithm, the objective function, and its optimization methods. In Section 4, this paper further introduces the parameter estimation based on the MM algorithm, including the application of the MM algorithm in anomaly detection and the specific implementation method of parameter estimation. In Section 5, the convergence and complexity of the algorithm are analyzed. In Section 6, the simulation results analysis of this paper are given, and the superiority and feasibility of the anomaly detection method based on the MM algorithm are proved through a detailed analysis of the experimental results. Finally, in Section 7, the paper is summarized.

2. System Model

2.1. Measurement Model

The measurement vector z is a nonlinear function [24,25] of the state vector $\mathbf{x}$ and can be expressed as

$$z = h(\mathbf{x}) + \boldsymbol{e} \tag{1}$$

where $\boldsymbol{e} = [e_1, e_2, \ldots, e_m]^T \in R^{m\times 1}$ is the measurement error vector, and $\boldsymbol{e}$ is assumed to be the Gaussian measurement noise with zero mean and covariance σ^2 , i.e., $\boldsymbol{e} \sim \mathcal{N}(0, \sigma^2)$. $h = [h_1(\mathrm{x}), h_2(\mathrm{x}), \ldots, h_m(\mathrm{x})]^T \in R^{m\times 1}$, $h_i(\mathbf{x})$ is a nonlinear function relating the ith measurement to the state vector $\mathbf{x}$. An alternative expression for the nonlinear relationship $h(\cdot)$ between the state vector and the measurement vector can be given by the following:

$$P_i = \mathrm{V}_i \sum_{j\in\Omega_i} \mathrm{V}_j (G_{ij}\cos\theta_{ij} + B_{ij}\sin\theta_{ij}) \tag{2}$$

$$Q_i = \mathrm{V}_i \sum_{j\in\Omega_i} \mathrm{V}_j (G_{ij}\sin\theta_{ij} - B_{ij}\cos\theta_{ij}) \tag{3}$$

$$P_{ij} = \mathrm{V}_i^2 (g_{si} + g_{ij}) - \mathrm{V}_i \mathrm{V}_j (g_{ij}\cos\theta_{ij} + b_{ij}\sin\theta_{ij}) \tag{4}$$

$$Q_{ij} = -\mathrm{V}_i^2 (b_{si} + b_{ij}) - \mathrm{V}_i \mathrm{V}_j (g_{ij}\sin\theta_{ij} - b_{ij}\cos\theta_{ij}) \tag{5}$$

In this expression, P_i and Q_i denote the active and reactive power injection at bus i, respectively; $P_{ij}$and Q_{ij} denote the real and reactive power flow from bus i to bus j, respectively; V_i represents the voltage at bus i; θ_i denotes the phase angle at bus i; θ_{ij} denotes the phase difference between buses i and j; $G_{ij} + jB_{ij}$ represents the line conductance between buses i and j; $g_{ij} + jb_{ij}$ represents the conductance of the branch branch of bus i; and Ω_i is the set of buses associated with bus i.

In power system SE, the state vector, $\mathbf{x}$, is usually composed of all nodes voltage amplitudes and their corresponding angles. The measurement vector, z, consists of active and reactive power injection and flow, voltage, and current magnitudes obtained from SCADA.

The estimation of the state vector $\mathbf{x}$ can be iteratively solved using weighted least squares (WLS); then,

$$\mathbf{x}^{k+1} = \mathbf{x}^k + [\boldsymbol{H}^{kT}\boldsymbol{R}^{-1}\boldsymbol{H}^k]^{-1}\boldsymbol{H}^{kT}\boldsymbol{R}^{-1}(z - h(\mathbf{x}^k)) \tag{6}$$

where $\mathbf{x}^k$ is the solution vector at the kth iteration; $\boldsymbol{R} = E[e \cdot e^T]$ is the measurement error covariance matrix; and $\boldsymbol{H} = (\partial h(\mathrm{x})/\partial \mathrm{x})$ is the measurement Jacobian matrix, that is

$$H(x) = \begin{bmatrix} \frac{\partial P_i}{\partial \theta} & \frac{\partial P_i}{\partial V} \\ \frac{\partial Q_i}{\partial \theta} & \frac{\partial Q_i}{\partial V} \\ \frac{\partial P_{ij}}{\partial \theta} & \frac{\partial P_{ij}}{\partial V} \\ \frac{\partial Q_{ij}}{\partial \theta} & \frac{\partial Q_{ij}}{\partial V} \end{bmatrix} \quad (7)$$

The weighted least squares iterative algorithm is a common method that is used to estimate state vectors.

2.2. Measurement Model with Outliers

If the observations completely conform to the measurement model in (1), the estimation $\hat{\mathbf{x}}$ can be obtained using WLS. However, when a few outliers exist in the measurement, the estimation performance may be seriously degraded. Outliers are deviations or anomalies in data that can be caused by a variety of factors, including sensor failures, communication or human errors, power equipment failures, damages, and errors in data transmission or processing. In the presence of outliers, the observations do not follow the model in (1) exactly, but they can obey the following model:

$$z = h(\mathbf{x}) + \boldsymbol{a} + \boldsymbol{e} \quad (8)$$

where the vector $\boldsymbol{a}$ denotes the outliers contained in the observations.

Traditional outlier detection methods usually use normalized residuals obtained based on weighted least squares to determine the presence of outliers in the data. However, this method has limitations in facing residual contamination and residual flooding problems. To overcome these problems, some improved outlier detection methods can be considered.

2.3. Gaussian Mixture Model for Measurements

Suppose that there are S outliers for M measurements of the measurement vector z. In the measurement acquisition process, we perform outlier detection by acquiring $L(L \geq 1)$ measurement vectors to determine whether there are outliers in the measurements. In the presence of outliers, the noise characteristics of the data are altered. Therefore, the ith measurement of the lth measurement vector z_l is denoted as follows:

$$z_{i,l} = \begin{cases} h_{i,l}(\mathbf{x}) + e_{i,l,1} \\ h_{i,l}(\mathbf{x}) + e_{i,l,2} \end{cases} \quad (9)$$

where $e_{i.l,1}$ and $e_{i.l,2}$ obey the Gaussian distribution, respectively, i.e., $e_{i.l,1} \sim \mathcal{N}(\mu_1, \sigma_1^2)$ and $e_{i.l,2} \sim \mathcal{N}(\mu_2, \sigma_2^2)$. $e_{i.l,1}$ denotes the measurement error vector in the absence of outliers. $e_{i.l,2}$ denotes the measurement error vector in the presence of outliers. Hence, the Gaussian mixture model (GMM) [26–28] is introduced to represent the probability density of $\boldsymbol{e}$, and $p(\boldsymbol{e})$ is given by

$$p(\boldsymbol{e}) = \sum_{k=1}^{2} \pi_k \mathcal{N}(\boldsymbol{e}|\mu_k, \sigma_k^2) \quad (10)$$

where $\mathcal{N}(\boldsymbol{e}|\mu_k, \sigma_k^2)$ denotes the kth component in the mixture model and π_k is the mixture coefficient and satisfies

$$\sum_{k=1}^{2} \pi_k = 1, 0 \leq \pi_k \leq 1 \quad (11)$$

It is assumed that M represents the number of measurements and S represents the number of outliers, hence $\pi_1 = (M-S)/M, \pi_2 = S/M$. The parameters that need to be estimated in GMM are $\boldsymbol{\theta} = [\pi_1, \pi_2, \mu_1, \mu_2, \sigma_1^2, \sigma_2^2]^T$. Then, the objective function of $\boldsymbol{\theta}$ can be written as

$$\begin{aligned} & J(\boldsymbol{\theta}, (\mathbf{z} - h(\mathbf{x}))) \\ & = \ln p((z - h(\mathbf{x})); \boldsymbol{\theta}) \\ & = \ln(\sum_{k=1}^{2} \pi_k \mathcal{N}((z - h(\mathbf{x}))|\mu_k, \sigma_k^2)) \\ & = \sum_{i=1}^{M} \sum_{l=1}^{L} \ln(\sum_{k=1}^{2} \pi_k \mathcal{N}((z_{i,l} - h_{i,l}(\mathbf{x}))|\mu_k, \sigma_k^2)) \end{aligned} \tag{12}$$

When unknown parameters are estimated, we cannot use the maximum likelihood method to derive the parameters that maximize the likelihood function as the single Gaussian model (SGM) does. To solve this problem, the Minorization–Maximization (MM) algorithm can be used. Through the MM algorithm, we can iteratively calculate the parameters in GMM.

3. The Minorization–Maximization Algorithm

For all observation data, it is not known in advance which sub-distribution they belong to. Each submodel has unknown parameters, and direct derivation cannot be calculated. It needs to be solved with an iterative approach. Therefore, the Minorization–Maximization algorithm can be used to solve the problem that unknown parameters are difficult to solve.

The MM algorithm is an iterative approach. The basic idea is to find a function that is easier to optimize as a surrogate function for the objective function and to indirectly optimize the objective function by optimizing the surrogate function. In each iteration, a new alternative function is constructed based on the parameter estimates obtained from the previous iteration. The new substitution function is then optimized to obtain the parameter estimate in this iteration and to use it in the calculation of the next iteration. Through continuous iteration, the estimated value of the parameter constantly approaches the optimal solution of the objective function.

According to the basic idea of MM algorithm, the surrogate function should satisfy

$$\begin{cases} Q(\boldsymbol{\theta}|\boldsymbol{\theta}^{(s)}) \le J(\boldsymbol{\theta}, (\mathbf{z} - h(\mathbf{x}))) \\ Q(\boldsymbol{\theta}^{(s)}|\boldsymbol{\theta}^{(s)}) = J(\boldsymbol{\theta}^{(s)}, (\mathbf{z} - h(\mathbf{x}))) \end{cases} \tag{13}$$

where $\boldsymbol{\theta}^{(s)}$ denotes the sth iteration of $\boldsymbol{\theta}$ and the surrogate function $Q(\boldsymbol{\theta}|\boldsymbol{\theta}^{(s)})$ is always below the objective function $J(\boldsymbol{\theta}, z)$. When $\boldsymbol{\theta} = \boldsymbol{\theta}^{(s)}$, the surrogate function $Q(\boldsymbol{\theta}|\boldsymbol{\theta}^{(s)})$ is tangent to the function $J(\boldsymbol{\theta}, z)$. Then, the surrogate function $Q(\boldsymbol{\theta}|\boldsymbol{\theta}^{(s)})$ is maximized to obtain

$$\boldsymbol{\theta}^{(s+1)} = \arg\max_{\boldsymbol{\theta}} \quad Q(\boldsymbol{\theta}|\boldsymbol{\theta}^{(s)}) \tag{14}$$

as the $(s+1)$th iteration of the $\boldsymbol{\theta}$. Finally, the maximum likelihood algorithm is used to estimate the unknown parameters.

From the basic principles of the MM algorithm, the difficulty of using the MM algorithm to estimate unknown parameters is in constructing a suitable function as a surrogate function of the objective function. The construction of the surrogate function is described in the next section.

4. Parameter Estimation of GMM

Constructing a suitable surrogate function is the key to estimating unknown parameters using the MM algorithm. Tian et al. [17] propose a new Assembly and Decomposition method (AD) to construct the surrogate function. Among them, technology is the basis of the MM algorithm, which guides the construction of the surrogate function. The D technique decomposes the objective function and then optimizes it.

4.1. Construction of Surrogate Function

According to the AD method, Jensen's inequality is used to construct a surrogate function in the minorization step. Jensen's inequality is given by the Equation (10), that is,

$$\varphi(\sum_{i=1}^{n} \alpha_i x_i) \geq \sum_{i=1}^{n} \alpha_i \varphi(x_i) \tag{15}$$

where we let $\varphi(\cdot)$ be a concave function, $\alpha_i \geq 0$, and $\sum\limits_{i=1}^{n} \alpha_i = 1$. Therefore, the surrogate function is expressed as

$$\begin{aligned} Q(\boldsymbol{\theta}|\boldsymbol{\theta}^{(s)}) &= \sum_{i=1}^{M}\sum_{l=1}^{L}[\sum_{k=1}^{2} w_{i,l,k} \ln \frac{\pi_k \mathcal{N}((z_{i,l} - h_{i,l}(\mathbf{x}))|\mu_k, \sigma_k^2)}{w_{i,l,k}}] \\ &= \sum_{i=1}^{M}\sum_{l=1}^{L}\sum_{k=1}^{2} w_{i,l,k} \ln(\pi_k \mathcal{N}((z_{i,l} - h_{i,l}(\mathbf{x}))|\mu_k, \sigma_k^2) + c^{(s)} \end{aligned} \tag{16}$$

where $\boldsymbol{\theta}^{(s)}$ represents the sth iteration of $\boldsymbol{\theta}$ and the weight function $w_{i,l,k}$ is denoted as

$$w_{i,l,k} = \frac{\pi_k^{(s)} \mathcal{N}((z_{i,l} - h_{i,l}(\mathbf{x}))|\mu_k^{(s)}, \sigma_k^{2,(s)})}{p((z_{i,l} - h_{i,l}(\mathbf{x})); \boldsymbol{\theta}^{(s)})} \tag{17}$$

where

$$p((z_{i,l} - h_{i,l}(x)); \boldsymbol{\theta}^{(s)}) = \sum_{k=1}^{2} \pi_k \mathcal{N}((z_{i,l} - h_{i,l}(x))|\mu_k^{(s)}, \sigma_k^{2,(s)}) \tag{18}$$

Define $\Phi_{i,l,k}^{(s)} = \pi_k^{(s)} \mathcal{N}((z_{i,l} - h_{i,l}(x))|\mu_k^{(s)}, \sigma_k^{2,(s)})$. By comparing the values of $\Phi_{i,l,k}^{(s)}$, the following weight function is given.

$$w_{i,l}^{(s)} = \begin{cases} 1, & \Phi_{i,l,1}^{(s)} \geq \Phi_{i,l,2}^{(s)} \\ 0, & \Phi_{i,l,1}^{(s)} < \Phi_{i,l,2}^{(s)} \end{cases} \tag{19}$$

The weight function satisfies

$$w_{i,l,k} \geq 0, \sum_{k=1}^{2} w_{i,l,k} = 1 \tag{20}$$

and

$$c^{(s)} = -\sum_{i-1}^{M}\sum_{l-1}^{L}\sum_{k-1}^{2} w_{i,l,k} \ln w_{i,l,k} \tag{21}$$

is a constant term independent of parameter $\boldsymbol{\theta}$.

4.2. Parameter Estimation of π_k, μ_k, and σ_k^2

The surrogate function in Equation (9) can be decomposed as

$$Q(\boldsymbol{\theta}|\boldsymbol{\theta}^{(s)}) = \sum_{i=1}^{M}\sum_{l=1}^{L}\sum_{k=1}^{2} w_{i,l,k} \ln \pi_k + \sum_{i=1}^{M}\sum_{l=1}^{L}\sum_{k=1}^{2} w_{i,l,k} \ln(\mathcal{N}((z_{i,l} - h_{i,l}(\mathbf{x}))|\mu_k, \sigma_k^2)) + c^{(s)} \tag{22}$$

where

$$J_1{}^{(s)}(\pi_k) = \sum_{i=1}^{M}\sum_{l=1}^{L}\sum_{k=1}^{2} w_{i,l,k} \ln \pi_k \tag{23}$$

$$J_2^{(s)}(\mu_k, \sigma_k^2) = \sum_{i=1}^{M}\sum_{l=1}^{L}\sum_{k=1}^{2} w_{i,l,k} \ln(\mathcal{N}((z_{i,l} - h_{i,l}(\mathbf{x}))|\mu_k, \sigma_k^2)) \tag{24}$$

In the $(s+1)$th iteration, the maximum likelihood estimation (MLE) of parameter $\boldsymbol{\theta}$ can be constructed by maximizing the surrogate function, that is

$$\boldsymbol{\theta}^{(s+1)} = \arg\max_{\boldsymbol{\theta}} \quad Q(\boldsymbol{\theta}|\boldsymbol{\theta}^{(s)}) \tag{25}$$

The expansion of Equation (24) can be written as

$$\begin{aligned} & J_2^{(s)}(\mu_k, \sigma_k^2) \\ & = \sum_{i=1}^{M}\sum_{l=1}^{L}[w_{i,l,1}\ln(\mathcal{N}((z_{i,l}-h_{i,l}(\mathbf{x}))|\mu_1,\sigma_1^2)) + w_{i,l,2}\ln(\mathcal{N}((z_{i,l}-h_{i,l}(\mathbf{x}))|\mu_2,\sigma_2^2))] \\ & = \sum_{i=1}^{M}\sum_{l=1}^{L}[w_{i,l,1}(\ln\frac{1}{(2\pi\sigma_1^2)^{\frac{N}{2}}} - \frac{(z_{i,l}-h_{i,l}(\mathbf{x})-\mu_1)^2}{2\sigma_1^2}) + w_{i,l,2}(\ln\frac{1}{(2\pi\sigma_2^2)^{\frac{N}{2}}} - \frac{(z_{i,l}-h_{i,l}(\mathbf{x})-\mu_2)^2}{2\sigma_2^2})] \end{aligned} \tag{26}$$

The partial derivative of the surrogate function with respect to unknown parameters can be obtained

$$\frac{\partial}{\partial \pi_k}[J_1^{(s)}(\pi_k)] = 0 \tag{27}$$

$$\frac{\partial}{\partial \mu_k}[J_2{}^{(s)}(\mu_k, \sigma_k^2)] = 0 \tag{28}$$

$$\frac{\partial}{\partial \sigma_k^2}[J_2{}^{(s)}(\mu_k, \sigma_k^2)] = 0 \tag{29}$$

and an iterative formula for the parameters can be obtained by solving the above equation, as follows:

$$\pi_k^{(s+1)} = \frac{\sum\limits_{i=1}^{M}\sum\limits_{l=1}^{L} w_{i,l,k}}{ML} \tag{30}$$

$$\mu_k^{(s+1)} = \frac{\sum\limits_{i=1}^{M}\sum\limits_{l=1}^{L} (z_{i,l} - h_{i,l}(\mathbf{x}^{(s)}))w_{i,l,k}}{\sum\limits_{i=1}^{M}\sum\limits_{l=1}^{L} w_{i,l,k}} \tag{31}$$

$$\sigma_k^{2,(s+1)} = \frac{\sum\limits_{i=1}^{M}\sum\limits_{l=1}^{L} (z_{i,l} - h_{i,l}(\mathbf{x}^{(s)}) - \mu_k^{(s+1)})^2 w_{i,l,k}}{\sum\limits_{i=1}^{M}\sum\limits_{l=1}^{L} w_{i,l,k}} \tag{32}$$

Then, The detailed workflow for estimating GMM parameters based on MM algorithm is shown in Algorithm 1.

Algorithm 1 GMM parameters are estimated based on the MM algorithm

Input: The measurements $\boldsymbol{z}$;

Initialize: Iteration index s=0 for MM algorithm;convergence tolerance is δ; and maximum iteration number is $N_{itr}^{\max}$.Given the initial value $\theta^{(0)} = [\pi_k^{(0)}, \mu_k^{(0)}, \sigma^{2(0)}]^T$

MM algorithm loop:

(1) Let $s = 1$ substitut the given initial value $\boldsymbol{\theta}^{(0)}$ into the iteration Formula (16) to calculate $\boldsymbol{\theta}^{(s)}$.

(2) Substituting $\boldsymbol{\theta}^{(s)}$ into the iteration formula (16) to calculate $\boldsymbol{\theta}^{(s+1)}$.;

(3) If the convergence condition $\left|J(\boldsymbol{\theta}^{(s+1)}, (\mathbf{z} - h(\mathbf{x}))) - J(\boldsymbol{\theta}^{(s)}, (\mathbf{z} - h(\mathbf{x})))\right| < \delta$ or $s + 1 = N_{itr}^{\max}$ is satisfied, the MM algorithm is terminated.

Output: $\left\{\pi_k^{(s+1)}, \mu_k{}^{(s+1)}, \sigma_k^{2,(s+1)}\right\}$.

5. Algorithm Analysis

5.1. Convergence Analysis

Since the MM algorithm is essentially iterative, the following inference is derived from the question of whether this algorithm can guarantee convergence [29].

The convergence of the algorithm refers to the monotonic convergence of the MM algorithm to some stationary point J^* of the log-likelihood function $J(\boldsymbol{\theta}, (z - h(\mathbf{x})))$. To prove this inference, we first need to prove the detailed workflow of the MM algorithm shown below.

$$J\left(\boldsymbol{\theta}^{(s+1)}; \boldsymbol{\theta}^{(s)}\right) \geq J\left(\boldsymbol{\theta}^{(s)}; \boldsymbol{\theta}^{(s)}\right) \tag{33}$$

holds for any $\boldsymbol{\theta}^{(s)}$ in its parameter space. The proof is as follows: For a given a priori position estimate $\mathbf{x}^{(s)}$, it is easy to show that updates $\pi_1^{(s+1)}, \pi_2^{(s+1)}, \mu_1{}^{(s+1)}, \mu_2{}^{(s+1)}, \sigma_1^{2,(s+1)}$ and $\sigma_2^{2,(s+1)}$ for the Gaussian distribution are global optimal solutions to the corresponding maxi- mization problems. Therefore, we can easily conclude that

$$J\left(\boldsymbol{\theta}_m^{(s+1)}, \mathbf{x}^{(s)}; \boldsymbol{\theta}^{(s)}\right) \geq J\left(\boldsymbol{\theta}_m^{(s)}, \mathbf{x}^{(s)}; \boldsymbol{\theta}^{(s)}\right) \tag{34}$$

where the right-hand side is exactly $J(\boldsymbol{\theta}^{(s)}; \boldsymbol{\theta}^{(s)})$.

The new estimate $\mathbf{x}^{(s+1)}$ is obtained by minimizing

$$f(x) = \sum_{i=1}^{N} \sum_{k=1}^{2} \frac{\left(z_i - h_i(\mathbf{x}) - \mu_k^{(s+1)}\right)^2}{\sigma_k^{2,(s+1)}} w_{ik} \tag{35}$$

using the BFGS quasi-Newton method with an initial guess set of $\mathbf{x}^{(s)}$. The BFGS quasi-Newton method [30–32] guarantees downhill progression toward the local minimum in the Newton step of each iteration, so that the new estimate $\boldsymbol{\theta}^{(s+1)}$ does not make $J(\boldsymbol{\theta}^{(s)}; \boldsymbol{\theta}^{(s)})$ decrease in the $(s+1)$th iteration, that is

$$J\left(\boldsymbol{\theta}_m^{(s+1)}, \mathbf{x}^{(s+1)}; \boldsymbol{\theta}^{(s)}\right) \geq J\left(\boldsymbol{\theta}_m^{(s+1)}, \mathbf{x}^{(s)}; \boldsymbol{\theta}^{(s)}\right) \tag{36}$$

where the left-hand side is identical to $J(\boldsymbol{\theta}^{(s+1)}; \boldsymbol{\theta}^{(s)})$. Thus, (36) can be proved. This means that the value of $J(\boldsymbol{\theta}, (z - h(\mathbf{x})))$ increases monotonically with iteration. Since it is bounded from above, convergence to some stationary point J^* of the $J(\boldsymbol{\theta}, (z - h(\mathbf{x})))$ is guaranteed.

5.2. Complexity Analysis

Complexity evaluation is often assessed using floating point operations (FLOPs) as a metric. Floating point arithmetic is a common metric for measuring the amount of computation required to execute an algorithm. It can be used to compare the computational complexity and efficiency of different algorithms. The computational complexity of the MM algorithm under the Gaussian mixture model is shown as follows. We will first define the floating point operations required for some basic operations.

(1) ε_{add}: FLOPs for addition.
(2) ε_{sub}: FLOPs for subtraction.
(3) ε_{mul}: FLOPs for multiplication.
(4) ε_{div}: FLOPs for division.
(5) ε_{exp}: FLOPs for exponents.
(6) ε_{pow}: FLOPs for raising to a real power.
(7) ε_{sqrt}: FLOPs for square roots.

Since the MM algorithm is an iterative algorithm, we analyze the $s+1$th iteration. Given a priori parameter estimation, the first step of the estimation is to evaluate the $w_{i,l,k}$ of $i = 1,2,\ldots,M, l = 1,2,\ldots,L$ and $k = 1,2$. This requires a calculation:

$$z_{i,l} - h_{i,l}(\mathbf{x}^{(s)}) \tag{37}$$

for $i = 1,2,\ldots,M, l = 1,2,\ldots,L$.

$$\phi_{i,l,k} = \frac{\pi_k^{(s)}}{\sqrt{2\pi\sigma_k^{2,(s)}}} \cdot \exp[\frac{-(z_{i,l} - h_{i,l}(x^{(s)}) - \mu_k^{(s)})^2}{2\sigma_k^{2,(s)}}] \tag{38}$$

for $i = 1,2,\ldots,M, l = 1,2,\ldots,L, k = 1,2$.

$$w_{i,l,k} = \frac{\pi_k^{(s)}\mathcal{N}((z_{i,l} - h_{i,l}(\mathbf{x}))|\mu_k^{(s)}, \sigma_k^{2,(s)})}{p((z_{i,l} - h_{i,l}(\mathbf{x}));\boldsymbol{\theta}^{(s)})} \tag{39}$$

for $i = 1,2,\ldots,M, l = 1,2,\ldots,L, k = 1,2$. Thus, Equation (37) requires $ML\varepsilon_{sub}$ FLOPs, Equation (38) requires $2((ML+3)\varepsilon_{mul} + (ML+1)\varepsilon_{pow} + ML\varepsilon_{sub} + ML\varepsilon_{div} + ML\varepsilon_{exp})$ FLOPs, and Equation (39) requires $ML(1\varepsilon_{sub} + 1\varepsilon_{add} + 1\varepsilon_{div})$ FLOPs. Then,

$$\pi_k^{(s+1)} = \frac{\sum_{i=1}^{M}\sum_{l=1}^{L} w_{i,l,k}}{ML} \tag{40}$$

$$\mu_k^{(s+1)} = \frac{\sum_{i=1}^{M}\sum_{l=1}^{L} (z_{i,l} - h_{i,l}(\mathbf{x}^{(s)}))w_{i,l,k}}{\sum_{i=1}^{M}\sum_{l=1}^{L} w_{i,l,k}} \tag{41}$$

$$\sigma_k^{2,(s+1)} = \frac{\sum_{i=1}^{M}\sum_{l=1}^{L} (z_{i,l} - h_{i,l}(\mathbf{x}^{(s)}) - \mu_k^{(s+1)})^2 w_{i,l,k}}{\sum_{i=1}^{M}\sum_{l=1}^{L} w_{i,l,k}} \tag{42}$$

for $i = 1,2,\ldots,M, l = 1,2,\ldots,L, k = 1,2$. It is easy to prove that Equation (40) requires $(NK-1)\varepsilon_{add} + 1\varepsilon_{div} + 1\varepsilon_{sub}$ FLOPs, Equation (41) requires $2(NK\varepsilon_{mul} + (NK-1)\varepsilon_{add} + \varepsilon_{div})$ FLOPs, and Equation (42) requires $2((NK+1)\varepsilon_{pow} + NK\varepsilon_{mul} + (NK-1)\varepsilon_{add} + \varepsilon_{div} + \varepsilon_{sub})$ FLOPs. Define $FL(\theta)$ as the total number of FLOPS consumed in one MM iteration to estimate θ. $FL(\theta)$ is equal to the total number of FLOPs consumed in Equation (37) through Equation (42), that is

$$\begin{aligned} FL(\theta) = &(6ML-5)\varepsilon_{add} + (4ML+3)\varepsilon_{sub} + (6ML+6)\varepsilon_{mul} \\ &+(4ML+4)\varepsilon_{pow} + (3ML+5)\varepsilon_{div} + 2ML\varepsilon_{\text{exp}} \end{aligned} \tag{43}$$

6. Simulation

To validate the feasibility of using the MM algorithm to detect outliers, this paper conducts a simulation analysis on the IEEE 14-bus system shown in Figure 1. Based on the relevant data in the Matpower power system simulation package, a conventional power flow calculation is performed, and the obtained system operation data are used as the measurement data of the power system. The simulation parameters utilized throughout the entire simulation process are summarized in Table 1, and the obtained experimental results are shown as follows.

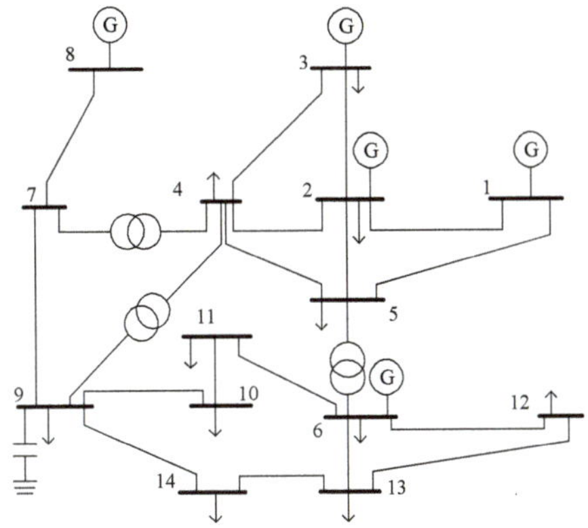

Figure 1. IEEE 14-bus system.

Table 1. Simulation parameters.

Parameter	Value
M	41
π_1	0.8
π_2	0.2
μ_1	0
μ_2	0.03
σ_1^2	0.01
σ_2^2	0.0025
Δ	10^{-6}
N_{itr}^{max}	100

In this study, simulation experiments containing 50 measurement vectors were analyzed. Figure 2 shows the distribution of the 2050 measurement errors. When there are outliers in the power system, part of the measurement error will change, assuming $|a_{i,l}| > 0$, where $a_{i,l}$ denotes the component of the outlier vector $\boldsymbol{a}$. Figure 3 shows the distribution of the measurement errors in the presence of outliers. Subsequently, the MM algorithm was used to classify the measurement errors and is presented in Figure 4.

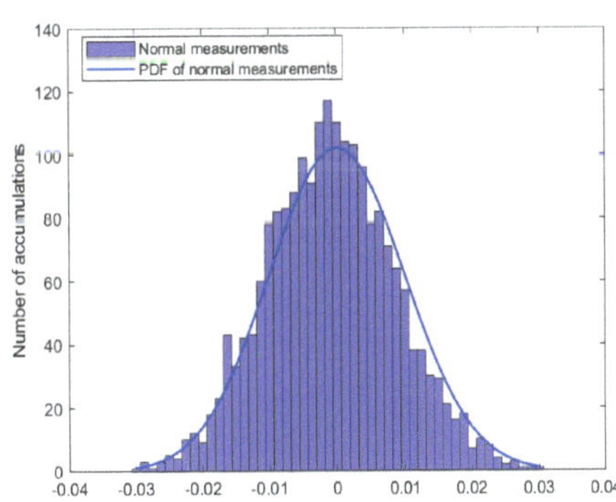

Figure 2. The distribution of measurements errors.

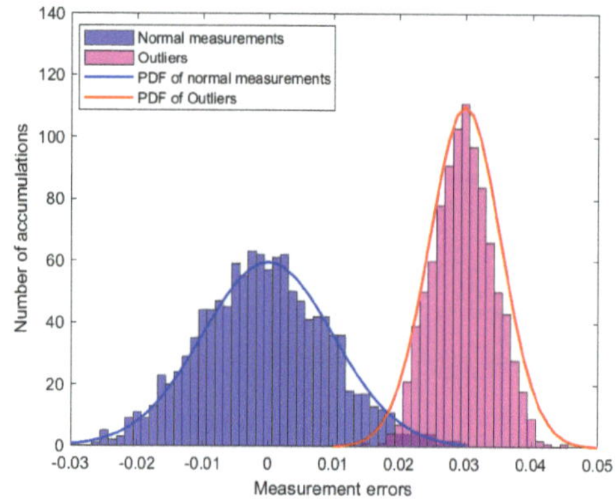

Figure 3. The distribution of measurement errors in the presence of outliers.

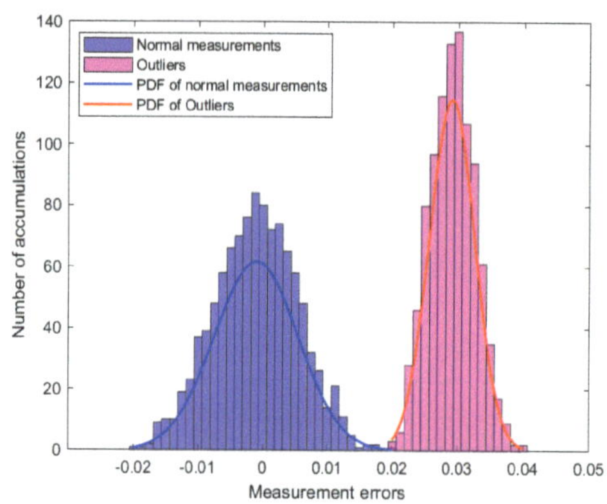

Figure 4. The distribution of measurements errors with MM algorithm.

To verify the convergence of the algorithm, it is assumed that M = 41, S = 8, and L = 50. The Monte Carlo method was used to perform 1000 independent experiments to detect the values of parameters $\boldsymbol{\theta} = [\pi_1, \pi_2, \mu_1, \mu_2, \sigma_1^2, \sigma_2^2]^T$. The errors between the mean of the estimated values and the actual values were calculated for each parameter. As shown in Figure 5, as the number of iterations increases, the estimated mean values of the parameters μ_2 and σ_1^2 gradually approach the true values, the error decreases, and the values finally converged to 0.00102 and 0.00134 at the sixth iteration, respectively. And, a variation in these errors was observed as the number of buses with outliers increased. The error between the mean and true values of the parameter estimates obtained from 1000 independent experiments was calculated. The equations are given by

$$E(\hat{\mu}_k) = \left| \frac{\sum\limits_{n=1}^{N} \hat{\mu}_{k,n}}{N} - \mu_k \right| \tag{44}$$

$$E(\hat{\sigma}_k^2) = \left| \frac{\sum\limits_{n=1}^{N} \hat{\sigma}_{k,n}^2}{N} - \sigma_k^2 \right| \tag{45}$$

and

$$E(\hat{\pi}_k) = \left| \frac{\sum_{n=1}^{N} \hat{\pi}_{k,n}}{N} - \pi_k \right| \tag{46}$$

$$E(\hat{\pi}_1) = \left| \frac{\sum_{n=1}^{N} \hat{\pi}_{1,n}}{N} - \pi_1 \right| \tag{47}$$

Since $\pi_1 > 0, \pi_2 > 0$ and $\pi_1 + \pi_2 = 1$, then the error for π_2 can be expressed as

$$E(\hat{\pi}_2) = \left| \pi_1 - \frac{\sum_{n=1}^{N} \hat{\pi}_{1,n}}{N} \right| \tag{48}$$

where $N = 1000$ and $\hat{\mu}_{k,n}$ denotes the estimate of $\hat{\mu}_k$ obtained in the nth experiment. The estimated mean is then obtained by summing $\hat{\mu}_{k,n}$ over 1000 independent experiments and dividing by N. And, due to $E(\pi_1) = E(\pi_2)$, the red line coincides with the blue line in Figure 6. As shown in Figures 6–8, it is evident that as the number of buses with outliers gradually increases, we clearly observe a gradual decrease in the error between the estimated mean and the actual values of the parameters π_2, μ_2 and σ_2. This is due to the increasing number of outliers as a proportion of the overall measurements. In this scenario, the estimated mean values of the parameters π_2, μ_2, and σ_2 gradually increase, getting closer to the true values, which leads to a gradual decrease in the error. In other words, as the number of outlier buses increases, we observe a significant improvement in the accuracy of the parameter estimates.

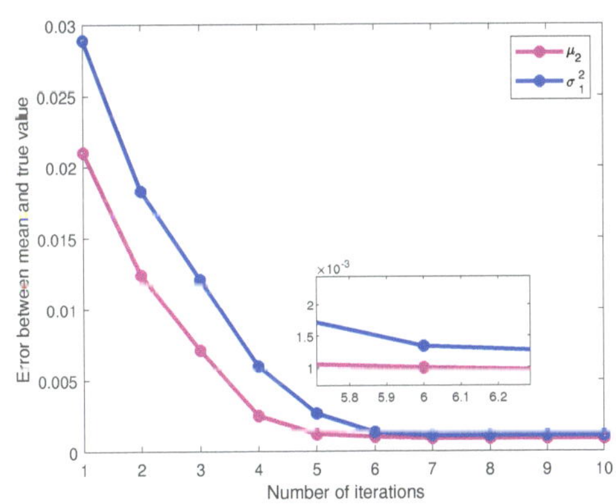

Figure 5. The error of the parameter varies with the number of iterations.

As shown in Figure 9, we assume that outliers exist in the power measurements of P_3, Q_3, P_{1-2}, P_{2-3}, P_{4-2}, Q_{1-2}, Q_{2-3}, and Q_{4-2} among the 41 measurements. To detect outliers, we conducted 1000 independent experiments and plotted Figure 10, showing the results of outlier detection based on the experimental results. By observing Figure 10, we can notice that the measurements in Q_3,P_{1-2},P_{4-2}, and Q_{4-2} are closer to the normal measurements. As a result, a small amount of data has been mistakenly identified as normal data. After applying the MM algorithm for outlier detection, we achieved an extremely high detection rate of over 95%. This implies that the vast majority of outlier values can be accurately detected.

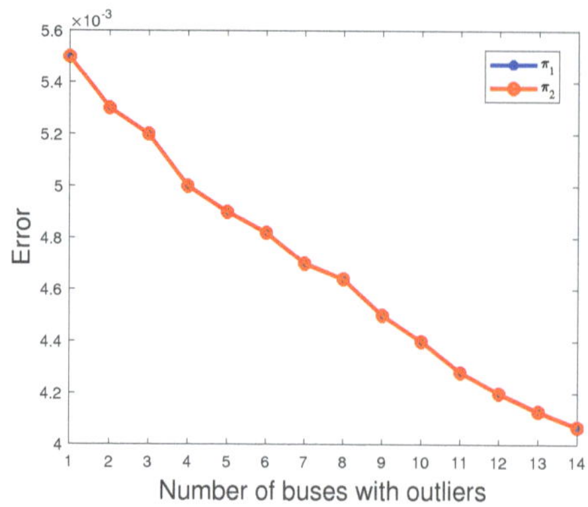

Figure 6. The error of parameter π_k varies with the number of buses with outliers.

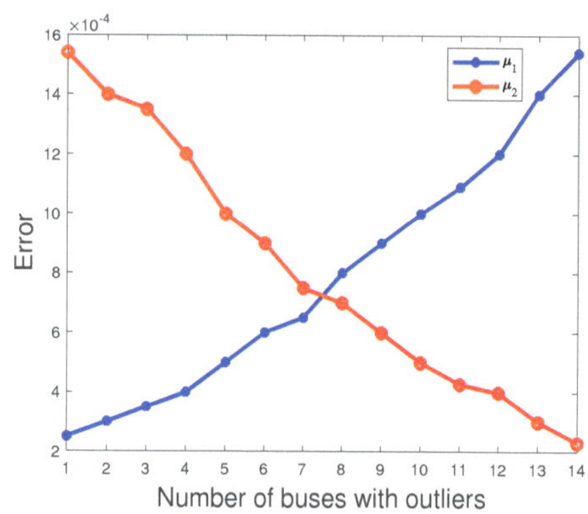

Figure 7. The error of parameter μ_k varies with the number of buses with outliers.

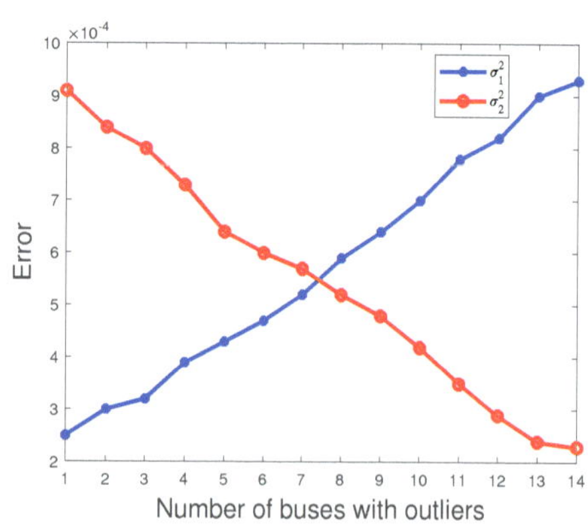

Figure 8. The error of parameter σ_k varies with the number of buses with outliers.

To validate the detection performance of the MM algorithm, WLS, and robust Z-score [33] for different outlier strengths, we use the detection rate as a performance metric.

Outlier strength is a metric used to quantify the degree of deviation of the outliers from the normal measurements. The outlier strength is defined as

$$\eta = \frac{\|a\|_2}{\|z\|_2} \tag{49}$$

where $a \in R^{m\times1}$ is the outlier vector and $z \in R^{m\times1}$ is the measurement vector. When the outlier strength is greater, this means that the deviation of the outlier relative to the normal measurement is greater and may have a more significant effect on the system. As shown in Figure 11, when the outlier strength is low, the outliers are not easily detected. However, the MM algorithm proposed in this paper outperforms WLS and robust Z-score in terms of detection performance. This is due to the iterative optimization process of the MM algorithm, which can gradually approach the optimal solution and shows good convergence, stability, and robustness in practice. Since the MM algorithm uses probabilistic models, it is better able to deal with outliers in the data and to reduce the impact of outliers on the results. The MM algorithm can limit the impact of outliers and provide more accurate parameter estimation, resulting in better performance in outlier detection. In contrast, WLS may suffer from outliers, leading to biased parameter estimates. The robust Z-score is more robust to outliers relative to the traditional Z-score, but MM algorithms are usually better at robustness to outliers in the modeling process. As the strength of the outliers varies, the false alarm rate also varies. As shown in Figure 12, the greater the strength of the outliers, the easier the outliers are detected and the smaller the false alarm rate. When the strength of the outliers is greater than 0.3, the false alarm rate decreases to less than 1%.

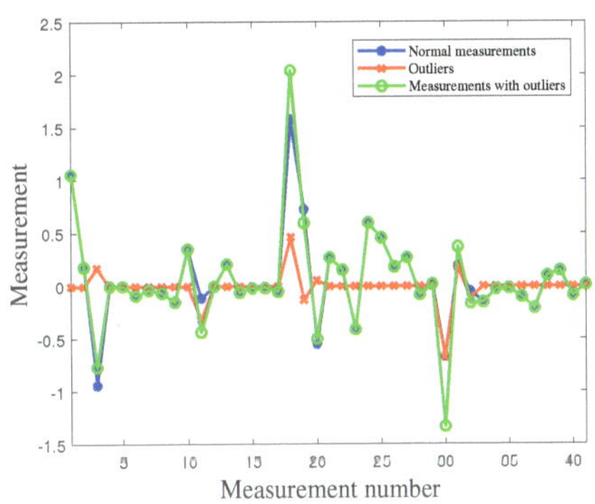

Figure 9. Data distribution before and after outliers in IEEE 14 bus system.

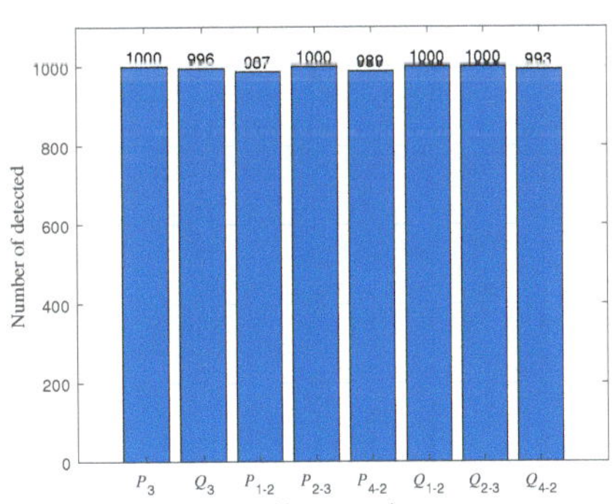

Figure 10. Outlier detection results.

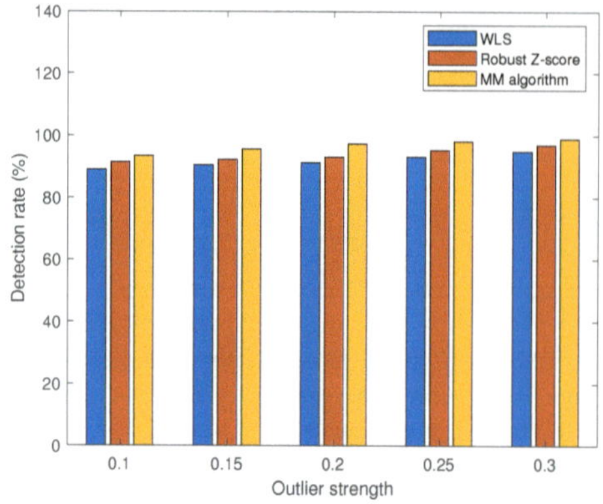

Figure 11. Comparison of detection performance at different outlier strengths.

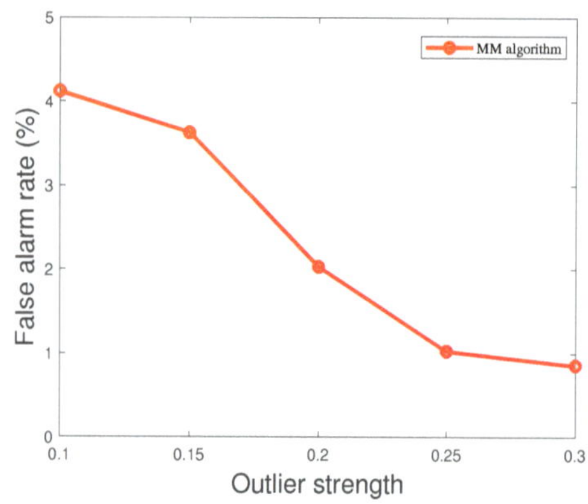

Figure 12. False alarm rate of MM algorithm for different outlier strengths.

7. Conclusions

Considering the impact of the presence of outliers on security in power systems, the introduction of MM algorithms for outlier detection and localization is an effective approach. The algorithm clusters and models the measurement errors using GMM and solves the model parameters iteratively using the MM algorithm. The feasibility and convergence of the MM algorithm in power systems are investigated through a simulation analysis on an IEEE 14-bus system, and the performance is compared with the WLS and robust Z-score at the same outlier intensity. The experimental results show that the MM algorithm can efficiently identify outliers in the power system and its detection rate can reach 95%. The results of this research emphasize the superiority of the MM algorithm in power system outlier detection. Compared with WLS and robust Z-score, the MM algorithm can model and detect outliers more accurately by introducing a probabilistic model and an iterative optimization algorithm to improve the security and reliability of the power system.

Author Contributions: Conceptualization, L.Q. and P.H.; methodology, Y.L. and W.G.; software, L.Q. and P.H.; validation, W.G.; formal analysis, F.H.; resources, W.G.; data curation, F.H.; writing—original draft preparation, L.Q.; writing—review and editing, W.G. and X.G.; visualization, L.Q. and P.H.; supervision, W.G.; project administration, Y.L.; funding acquisition, W.G. All authors have read and agreed to the published version of the manuscript.

Funding: This research was funded by the National Natural Science Foundation of China (NSFC) (U21A20146), Open Research Fund of Anhui Province Key Laboratory of Detection Technology and Energy Saving Devices (JCKJ2022A10), Collaborative Innovation Project of Anhui Universities (GXXT-2020-070), and Open Research Fund of Anhui Province Key Laboratory of Electric Drive and Control (DQKJ202103).

Institutional Review Board Statement: Not applicable.

Informed Consent Statement: Not applicable.

Data Availability Statement: Not applicable.

Acknowledgments: We thank the anonymous reviewers for their valuable comments.

Conflicts of Interest: The authors declare no conflict of interest.

References

1. Zhao, J.; Gómez-Expósito, A.; Netto, M.; Mili, L.; Abur, A.; Terzija, V.; Kamwa, I.; Pal, B.; Singh, A.K.; Qi, J.; et al. Power system dynamic state estimation: Motivations, definitions, methodologies, and future work. *IEEE Trans. Power Syst.* **2019**, *34*, 3188–3198. [CrossRef]
2. Kang, J.W.; Xie, L.; Choi, D.H. Impact of data quality in home energy management system on distribution system state estimation. *IEEE Access* **2018**, *6*, 11024–11037. [CrossRef]
3. Wang, Y.; Tian, J.; Sun, Z.; Wang, L.; Xu, R.; Li, M.; Chen, Z. A comprehensive review of battery modeling and state estimation approaches for advanced battery management systems. *Renew. Sustain. Energy* **2020**, *131*, 110015. [CrossRef]
4. Zhao, J.; Mili, L. A theoretical framework of robust H-infinity unscented Kalman filter and its application to power system dynamic state estimation. *IEEE Trans. Signal Process.* **2019**, *67*, 2734–2746. [CrossRef]
5. Liu, S.; Zhao, Y.; Lin, Z.; Liu, Y.; Ding, Y.; Yang, L.; Yi, S. Data-driven event detection of power systems based on unequal-interval reduction of PMU data and local outlier factor. *IEEE Trans. Smart Grid* **2019**, *11*, 1630–1643. [CrossRef]
6. Wei, S.; Xu, J.; Wu, Z.; Hu, Q.; Yu, X. A False Data Injection Attack Detection Strategy for Unbalanced Distribution Networks State Estimation. *IEEE Trans. Smart Grid* **2023** , *14*, 3992–4006. [CrossRef]
7. Abdolkarimzadeh, M.; Aghdam, F.H. A novel and efficient power system state estimation algorithm based on Weighted Least Square (WLS) approach service. *J. Power Technol.* **2019**, *99*, 15–24. Available online: https://papers.itc.pw.edu.pl/index.php/JPT/article/view/1258 (accessed on 15 June 2023).
8. Wang, L.; Zhou, Q.; Jin, S. Physics-guided deep learning for power system state estimation. *J. Mod. Power Syst. Clean Energy* **2020**, *8*, 607–615. [CrossRef]
9. Kotha, S.K.; Rajpathak, B. Power system state estimation using non-iterative weighted least square method based on wide area measurements with maximum redundancy. *Electr. Pow. Syst. Res.* **2022**, *206*, 107794. [CrossRef]
10. Tan, S.; Wu, W.; Shao, Z.; Li, Q.; Li, B.; Huang, J. CALPA-NET: Channel-pruning-assisted deep residual network for steganalysis of digital images. *IEEE Trans. Inf. Forensics Secur.* **2020**, *16*, 131–146. [CrossRef]
11. Kirchgässner, W.; Wallscheid, O.; Böcker, J. Estimating electric motor temperatures with deep residual machine learning. *IEEE Trans. Power Electron.* **2020**, *36*, 7480–7488. [CrossRef]
12. Bhatti, B.A.; Broadwater, R. Distributed Nash equilibrium seeking for a dynamic micro-grid energy trading game with non-quadratic payoffs. *Energy* **2020**, *202*, 117709. [CrossRef]
13. Asadi, S.; Khayatian, A.; Dehghani, M.; Vafam, N.; Khooban, M.H. Robust sliding mode observer design for simultaneous fault reconstruction in perturbed Takagi-Sugeno fuzzy systems using non-quadratic stability analysis. *J. Vib. Control* **2020**, *26*, 1092–1105. [CrossRef]
14. Sun, Y.; Babu, P.; Palomar, D.P. Majorization-minimization algorithms in signal processing, communications, and machine learning. *IEEE Trans. Signal Process.* **2016**, *65*, 794–816. [CrossRef]
15. Arora, A.; Tsinos, C.G.; Rao, B.S.M.R.; Chatzinotas, S.; Ottersten, B. Hybrid transceivers design for large-scale antenna arrays using majorization-minimization algorithms. *IEEE Trans. Signal Process.* **2019**, *68*, 701–714. [CrossRef]
16. Panwar, K.; Katwe, M.; Babu, P.; Ghare, P.; Singh, K. A majorization-minimization algorithm for hybrid TOA-RSS based localization in NLOS environment. *IEEE Commun. Lett.* **2022**, *26*, 1017–1021. [CrossRef]
17. Tian, G.L.; Huang, X.F.; Xu, J. An assembly and decomposition approach for constructing separable minorizing functions in a class of MM algorithms. *Stat. Sin.* **2019**, *29*, 961–982. [CrossRef]
18. Tian, G.L.; Ju, D.; Chuen Yuen, K.; Zhang, C. New expectation–maximization-type algorithms via stochastic representation for the analysis of truncated normal data with applications in biomedicine. *Stat. Methods Med. Res.* **2018**, *27*, 2459–2477. [CrossRef]
19. Tavakoli, S.; Yooseph, S. Learning a mixture of microbial networks using minorization–maximization. *Bioinformatics* **2019**, *35*, i23–i30. [CrossRef]
20. Shen, K.; Yu, W.; Zhao, L.; Palomar, D.P. Optimization of MIMO device-to-device networks via matrix fractional programming: A minorization–maximization approach. *IEEE/ACM Trans. Netw.* **2019**, *27*, 2164–2177. [CrossRef]

21. Ortega, J.M.; Rheinboldt, W.C. *Iterative Solution of Nonlinear Equations in Several Variables*; Academic Press: New York, NY, USA, 1970.
22. De Leeuw, J. Applications of Convex Analysis to Multidimensional Scaling. 2005. Available online: https://escholarship.org/uc/item/7wg0k7xq (accessed on 6 July 2023).
23. Lange, K.; Hunter, D.R.; Yang, I. Optimization transfer using surrogate objective functions. *J. Comput. Graph. Stat.* **2000**, *9*, 1–20. Available online: https://www.tandfonline.com/doi/abs/10.1080/10618600.2000.10474858 (accessed on 3 July 2023).
24. Zhao, J.; Mili, L. A framework for robust hybrid state estimation with unknown measurement noise statistics. *IEEE Trans. Industr. Inform.* **2017**, *14*, 1866–1875. [CrossRef]
25. Zhang, T.; Wang, Y.; Ning, J.; Zhai, M. Detection and identification of bad data based on neural network and k-means clustering. In Proceedings of the IEEE Innovative Smart Grid Technologies-Asia (ISGT Asia), Chengdu, China, 21–24 May 2019; pp. 3634–3639. [CrossRef]
26. Huang, T.; Peng, H.; Zhang, K. Model selection for Gaussian mixture models. *Stat. Sin.* **2017**, *27*, 147–169. Available online: https://www.jstor.org/stable/44114365 (accessed on 5 May 2023). [CrossRef]
27. Hosseini, R.; Sra, S. An alternative to EM for Gaussian mixture models: Batch and stochastic Riemannian optimization. *Math. Program.* **2020**, *181*, 187–223. Available online: https://archive.ics.uci.edu/ml/datasets (accessed on 2 June 2023). [CrossRef]
28. Pfeifer, T.; Protzel, P. Expectation-maximization for adaptive mixture models in graph optimization. International conference on robotics and automation (ICRA). In Proceedings of the 2019 International Conference on Robotics and Automation (ICRA), Montreal, QC, Canada, 20–24 May 2019; pp. 3151–3157. [CrossRef]
29. Yin, F.; Fritsche, C.; Gustafsson, F.; Zoubir, A.M. EM-and JMAP-ML based joint estimation algorithms for robust wireless geolocation in mixed LOS/NLOS environments. *IEEE Trans. Signal Process.* **2013**, *62*, 168–182. [CrossRef]
30. Eisen, M.; Mokhtari, A.; Ribeiro, A. Decentralized quasi-Newton methods. *IEEE Trans. Signal Process.* **2017**, *65*, 2613–2628. [CrossRef]
31. Li, M. A three term Polak-Ribière-Polyak conjugate gradient method close to the memoryless BFGS quasi-Newton method. *J. Ind. Manag. Optim.* **2018**, *16*, 245–260. [CrossRef]
32. Li, M. A modified Hestense–Stiefel conjugate gradient method close to the memoryless BFGS quasi-Newton method. *Optim. Methods Softw.* **2018**, *33*, 336–353. [CrossRef]
33. Degirmenci, A.; Karal, O. Robust incremental outlier detection approach based on a new metric in data streams. *IEEE Access* **2021**, *9*, 160347–160360. [CrossRef]

sensors

MDPI

Article

Exploitation Perspective Index as a Support of the Management of the Transformer Fleet

Michał Kunicki [1,*], Sebastian Borucki [1] and Jan Fulneček [2]

[1] Department of Electrical Power Engineering and Renewable Energy, Opole University of Technology, 45-758 Opole, Poland; weai@po.edu.pl
[2] Department of Electrical Power Engineering, VSB Technical University of Ostrava, 708 00 Ostrava, Czech Republic; jan.fulnecek@vsb.cz
* Correspondence: m.kunicki@po.edu.pl

Abstract: This paper presents an alternative approach to the Transformer Assessment Index (TAI) by proposing a relatively simple rating method called the Exploitation Perspective Index (EPI). The method provides two numerical indicators: the first reflects the overall technical condition of the particular unit, and the second shows the condition of the unit in the context of the entire fleet. The objective of the EPI method is to support the decision-making process regarding the technical condition assessment of each of the transformers in the target population, considering not only technical but also economic aspects of transformer maintenance. Application of the method is described step by step, including input data, parametrization of the weights, and interpretation of the output results it provides. The proposed method is evaluated by two representative use cases and compared with two other methods. As a result, EPI confirms its applicability, and it has already been successfully implemented by the electric power industry. EPI can be potentially freely adopted for any transformer fleet, as well as for the specific situation of the utility, by adjusting the relevant parameters.

Keywords: power transformer; high voltage; measurements; condition assessment; fault diagnostics

Citation: Kunicki, M.; Borucki, S.; Fulneček, J. Exploitation Perspective Index as a Support of the Management of the Transformer Fleet. *Sensors* **2023**, *23*, 8681. https://doi.org/10.3390/s23218681

Academic Editor: Manuel Pineda-Sanchez

Received: 9 August 2023
Revised: 7 October 2023
Accepted: 23 October 2023
Published: 24 October 2023

1. Introduction

One of the priorities for utilities is to provide uninterrupted power supply to end users. The reliability of the electric power asset mainly depends on adequate maintenance and diagnostics [1–4]. Power transformers are no doubt one of the key elements of the entire power system, so their technical condition, especially faults, may directly affect the reliability of the system. Condition assessment of transformers is based on various measurement data, usually aimed at assessment of the insulation system and mechanical condition [5–8], as well as the influence of physical factors like temperature [9–11], load, and even weather [12]. Knowing the technical condition of a particular unit is only the first step in the maintenance and risk management chain regarding the entire fleet. Adequate management of the transformer fleet requires additional decision support tools, which indicate the absolute technical condition of a given unit in the context of the entire fleet. Such tools are called the Transformer Assessment Index (TAI) and may be implemented in various ways, briefly discussed below [13].

In [14], the authors considered the unavailability of the data as one of the limitations of TAI. In their opinion, such a situation may result in inadequate condition assessment of a transformer due to limited data calculation. To solve this problem, the authors proposed the application of parameter prioritization results to estimate the certainty level caused by the data availability of a TAI. To demonstrate the proposed, method five scenarios were used to calculate the certainty level. Validation of the proposed approach in the context of its influence on TAI was performed on a 150 kV power transformer with six points of measurement data. This problem was also raised by the same authors in [15], where a comparison of seven models to substitute unavailable data on paper insulation condition was presented. The health index of 200 transformers with complete data was calculated, and compared to the alternative models. The analysis showed that the optimized models were based on multiple linear regression and an adaptive neuro-fuzzy inference system. The problem of adjusting the weighting factor of TAI was raised in [16]. According to the authors, one of the possible solutions to this problem is the involvement of many experts, but usually such an approach leads to complexity in aggregating the results. As a result, a novel method to implement the consensus for multiple experts in transformer assessment index weighting factor de-termination based on the analytic hierarchy process was proposed. Aman et al. in [17] proposed a novel method for assessing the overall health condition of the transformer. A combination of the health index and the criticality index was used to prioritize the strategic decisions on the transformer. The proposed method divided the population into four groups, taking into account diagnostic indicators and the critical operations of individual transformers. Consideration of the transformer's apparent and actual age and their potential influence on the assessment of its aging condition was presented in [18]. The authors calculated the health index for 130 units and used them to model the curve of the heath index decrease in the entire population. The apparent age and aging condition of the transformer were then estimated using the proposed curve. The apparent age problem also appeared in [19], where a novel method for assessing the health index of transformers, considering the apparent age strategy, was proposed. The method used four input factors for the evaluation process (DGA, oil quality, bushings conditions, and equipment degradation). Apparent age was included in the method mainly by hot spot temperature and load data. In [20], the authors investigated the improvement of the TAI by using fuzzy logic methods. The integrated fuzzy model was developed by combining the sub-models of fuzzy logic. The proposed solution yielded an improved assessment of the transformer insulation condition. According to the authors, this approach decreases the complexity of life estimation and health index evaluation of power transformers. Another modeling approach regarding TAI was proposed in [21]. Zeinoddini-Meymand et al. analyzed the application of linear and nonlinear models to evaluate TAI. The authors compared two nonlinear models (artificial neural network and adaptive neuro-fuzzy inference system) with a multiple linear regression linear statistical model. Furthermore, they proposed an extended data set to optimize the TAI calculation, including 15 parameters (i.e., DGA and various oil properties). This study showed that an adaptive neuro-fuzzy inference system-based model provided the best results. A newly developed TAI was also proposed in [22]. A hierarchical health index model was established based on statistical product and service solutions. The proposed method takes into account the perspectives of transformer thermal, electrical, mechanical, and load. An interesting aspect was raised by Benhmed et al. in [23], where the authors investigated feature selection and classification techniques to reduce the complexities of TAI. Several filters and wrapper-based feature selection methods were investigated. The performance of the proposed approach is validated by evaluations of selected classification models. According to the authors, their method reduced the optimum number of features by separating only the most influential ones when calculating TAI. Furthermore, results showed that water content, acidity, breakdown voltage, and furans were the most influential testing parameters in calculating TAI. A novel health index, risk, and remaining lifetime estimation method for power transformers is discussed in [24]. The method was based on a combination of three selected models: a winding degradation physical model, a health index model based on condition monitoring data combined with expert judgment, and a statistics-based end-of-life model. Data from real-life transformers were used to validate the method. According to the authors, this method allows for the identification of transformers in poor condition and the follow-up and prioritization of transformers for maintenance and replacement.

Some of the current studies on TAI, apart from the assessment of the technical condition, also take into account economic aspects. In [25], the authors proposed an innovative approach to the maintenance decision-making model, considering reliability and cost-effectiveness. This approach was based on a particle swarm optimization method, which was used to optimize the proposed model and select the best maintenance strategy. The performance of the proposed method was verified in two cases, which confirmed a significant improvement in the maintenance strategy with this method. In [26], the authors used a health index to assess the maintenance cost of transformers. This method is based on the Markov model, which is used to predict the future state of the transformer. Future-state distribution probabilities were used to estimate the maintenance cost of the selected unit, according to the proposed maintenance policy model.

Further discussion and comparison of various contemporary approaches to TAI can be found in [13,27,28].

Thus, this paper proposes an alternative technical condition index method for power transformers called Exploitation Perspective Index (EPI). The objective of this method is to provide a simple, universal, and personalized tool to provide a brief view of the entire fleet as well as a particular transformer. The most relevant novelty regarding this method is that it takes into account both the technical condition of the assessed transformer and the economic aspect of the required maintenance. Furthermore, EPI provides two rating scales in one rating procedure: absolute and relative. The absolute scale corresponds with the current technical condition of the unit and its potential future exploitation perspective, while the relative scale reflects the overall technical condition of the particular unit in the context of the entire fleet. Finally, potentially, EPI can be freely adopted for any transformer fleet as well as for the specific situation of the utility by adjusting the relevant parameters.

2. Proposed EPI Method

2.1. Requirements for the Proposed Method

This section raises some key assumptions the proposed EPI method should meet regarding its functionality and range of applications. Most of the assumptions for EPI are derived from typical requirements that similar methods should comply with. Most of these requirements are grounded in applicable industry standards and common good practices in the field of technical condition assessment of power transformers. However, it must be emphasized that the EPI method is designed for a specific target transformer fleet, and as a result, some of the requirements and proposed assumptions may apply strictly to this fleet.

- The objective of the EPI method is to support the decision-making process regarding the technical condition assessment of each transformer in the target population;
- The method should yield a simple rating to show the overall technical condition of the particular unit in the context of the entire fleet;
- The method should use the conclusions from periodical routine test results (instead of raw measurement data) of the transformers typically performed in the fleet the EPI is designed for;
- EPI should not analyze any of the raw measurement data, while it should rather use an expert diagnosis (defects and other malfunctions detected on the grounds of the routine tests) for further analysis;
- EPI should be a numerical value that corresponds with the current technical condition of the unit and its potential future exploitation perspective (absolute rating scale);
- EPI should focus not only on technical but also economic aspects of transformer maintenance;
- EPI should also reflect the overall technical condition and future exploitation perspective of the particular unit in the context of the entire fleet (relative rating scale).

2.2. Description of the Proposed Method

2.2.1. Input Data Initial Preparation

One of the crucial steps prior to the implementation of EPI is the initial preparation of the input data. All of the inputs should be taken from the periodic transformer test reports. It should be emphasized that EPI does not analyze any raw measurement data, so it does not need any new technical rating criteria or testing procedures to be defined. It is assumed that experts or decision support systems identify potential defects and draw auxiliary conclusions (based on the analysis of raw measurement data). Therefore, to prepare input data for EPI, appropriate standardization of the transformer test report is required. To address this point, it is proposed to include a predefined checklist in the conclusion section of every report. The idea is not to influence any test procedures or diagnostic criteria, but only to present the final conclusions in a categorized and predefined way. According to this proposal, the supplementary task for the expert (or system) diagnosing the analyzed transformer (in the test report) would be to mark the relevant defects/conclusions in the attached checklist (simple binary criterion): mark it if applicable, and not to mark it if not applicable (based on the analysis of the measurement results). Definitions and interpretations of the proposed input parameters are provided in Table 1. This list contains most (or even all) of the typical defects that may be identified in the transformer.

Table 1. Description of the input parameters.

	Input Parameter	Definition
1	Minor oil leaks	Visual inspection indicated minor oil leaks in the transformer (other than 2, 11, 16, 17)—not relevant from the exploitation point of view
2	Major oil leaks	Visual inspection indicated major oil leaks relevant from the exploitation point of view (main tank, primary seal)
3	Minor paint loss or corrosion	Visual inspection indicated minor paint loss or corrosion, not relevant from the exploitation point of view
4	Major paint loss or corrosion	Visual inspection indicated major paint loss or corrosion, relevant from the exploitation point of view
5	Damage to the thermometer or invalid readings	Incorrect readings or damage to the upper oil layer thermometer were found
6	low oil level	Too low an oil level in the transformer conservator, below the permissible level
7	Damage to the oil level gauge or invalid readings	Incorrect, illegible readings, or damage to the oil level gauge
8	Buchholz relay fault	Damage or leaks or damage to the cables or lack of oil in the gas-flow relay (Buchholz)
9	Cooling system malfunction	Abnormalities in the operation of the cooling system (radiators, fans, control cabinet), other than 16
10	Grounding connection faults	Abnormalities in the connection and grounding of the transformer
11	Desiccant faults (dehydrating breather)	Abnormalities in the dehydrating system (leakage, moisture in the cartridge)
12	OLTC drive malfunction	Abnormalities in the operation of the PPZ drive
13	Bushing's damage	Visual inspection indicated mechanical damage to the bushing (other than 17)
14	Signaling and controlling wiring faults	Damage to the transformer's secondary and control circuits
15	Malfunction of the fiber optic temperature measurement system	Damage/abnormalities in the operation of the fiber-optic temperature measurement system of the active part of the transformer
16	Oil leaks (cooling system)	Oil leaks from radiators, pumps, valves or other components of the transformer cooling system
17	Oil leaks (bushings)	Oil leaks from bushings or their measuring taps
18	Moisture in the oil	Level of moisture in oil exceeded the allowed level
19	Aged oil	Aging markers of the oil indicate reaching the end of life or advanced aging process

Table 1. *Cont.*

	Input Parameter	Definition
20	Partial discharges	DGA results indicate PD
21	Overheating	DGA results and/or fiber optic temperature measurement results indicate overheating
22	Stray gassing	DGA results indicate stray gasses
23	Aged cellulose	Aging markers of the cellulose insulation indicate reaching the end of life or advanced aging process
24	Moisture in solid insulation	Level of moisture in solid insulation exceeded the allowed level
25	Aged bushings	Aging markers of the bushings indicate reaching the end of life or advanced aging process
26	Windings deformation	SFRA results indicate deformation of windings
27	Turn-to-turn short-circuits	Test results indicate turn-to-turn short-circuits
28	Windings asymmetry	Test results indicate winding asymmetry
29	Winding discontinuity	Test results indicate winding discontinuity
30	OLTC defects	OLTC time of non-simultaneous operation and/or head's own time exceed the criteria values, and/or discontinuity on any tap detected
31	Magnetic circuit defect	Test results indicate a defect in the magnetic circuit

2.2.2. Implementation of EPI

EPI is a relatively simple, dimensionless rating method for transformers that considers not only their technical condition but also an economic aspect of the required maintenance. The proposed rating procedure is based on the results of actual periodic routine tests of the transformer. EPI uses 2 types of data as input parameters (Table 2): defects (DEF) and auxiliary conclusions (AC). Furthermore, it also considers the transformer age. As described in Section 2.2.1, EPI does not analyze any raw measurement data, instead using conclusions from the periodic transformer test report as input. It is quite different from other methods discussed in Section 1 and makes this method unique in this regard. The procedure for determining the EPI consists of summing up the weights assigned to all DEF and AC, respectively, indicated for a given unit (Table 2). It follows that theoretically, the minimum possible value of the EPI is equal to 0—in a situation where an assessed unit has no identified defects and failures, and thus no maintenance recommendation has been generated for it. Therefore, the maximum value of the EPI is theoretically limited by the sum of the weights of all DEF and AC (but it is a theoretical case rather than a real-life scenario). A general flowchart of the EPI method is presented in Figure 1.

A general formula to calculate the EPI for a selected transformer *tr* is presented in (1):

$$\mathrm{EPI}_{tr} = \sum_{i} D_i, \quad (1)$$

where

$$D_i = \left(d_1 \cdot \frac{k_i \cdot m_i}{S_{tr}} + d_2 \cdot n_i \cdot P_{tr} \right) \cdot 100, \quad (2)$$

is a single weight related to a specific *i*-th input parameter of the analyzed transformer, according to Table 1. Equation (2) consists of two fundamental parts related to the economical and technical aspects of the assessed unit, respectively. These formulas are empirically based on the experience of authors and industry experts; they also correspond to industry standards and recommendations in this regard [13].

Table 2. Complete catalog of all possible input parameters for EPI and their proposed default values, D_i.

i	Input Parameter AC	D_i	i	Input Parameter DEF	D_i
1	minor oil leaks (not within the main tank)	1	18	moisture in the oil	6.0
2	major oil leaks (main tank, primary seal)	2.0	19	aged oil	7.0
3	minor paint loss or corrosion	1.0	20	partial discharges (based on DGA)	7.0
4	major paint loss or corrosion	2.0	21	overheating	5.0
5	damage to the thermometer or invalid readings	0.1	22	stray gassing	0.2
6	low oil level	1.0	23	aged cellulose	60.0
7	damage to the oil level gauge or invalid readings	1.5	24	moisture in solid insulation	20.0
8	Buchholz relay fault	0.4	25	aged bushings	5.0
9	cooling system malfunction	2.0	26	windings deformation	25.0
10	grounding connection faults	0.5	27	turn-to-turn short-circuits	50.0
11	desiccant faults (dehydrating breather)	0.1	28	windings asymmetry	3.0
12	OLTC drive malfunction	2.5	29	winding discontinuity	6.0
13	bushing's damage (visual)	3.0	30	OLTC defects	2.0
14	signaling and controlling wiring faults	2.5	31	magnetic circuit defect	38.0
15	malfunction of the fiber optic temperature measurement system	2.0			
16	oil leaks (cooling system)	1.0			
17	oil leaks (bushings)	1.0			

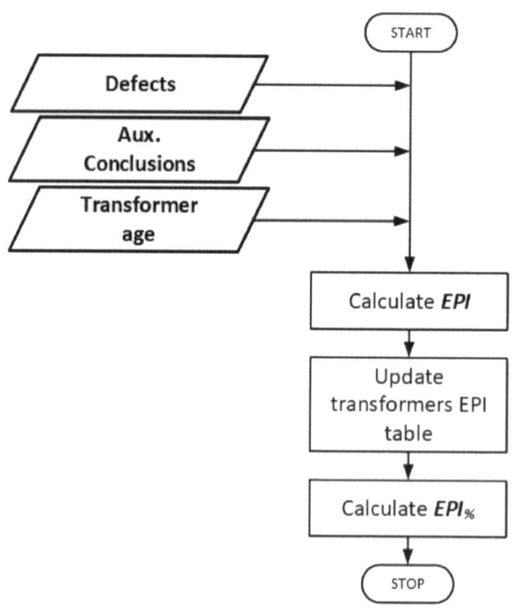

Figure 1. General flowchart of the EPI method.

First, the economical-related part uses d_1, k_i, m_i, and S_{tr} for calculation. S_{tr} is an estimated value (price) of the new transformer meeting the same requirements as the assessed one (equivalent). Despite S_{tr} having the same value for all input parameters (within the same assessment process), it may change over time, so it should be updated regularly. Estimated repair/replace/maintenance costs related to the i-th parameter are defined as m_i. This parameter has different values for every input parameter, as they require different maintenance procedures (and consequently costs). Its value may also vary over time, so it should be updated if necessary. Both S_{tr} and m_i should be expressed in the same currency. k_i can be defined as the economical weight of the i-th input parameter. Primarily, it should consider the age of the transformer (its depreciation), as usually some extended overhaul of the old units may not be rational from an economic point of view. Furthermore, k_i can also consider some other aspects that may be important from an economic point of view, e.g., possible transport costs of the transformer, availability of specific components at a given moment, etc. The default value of k_i for this EPI method is 1. In order to include the age of the transformer, one of the commonly proposed normalized aging (or loss of life, etc.) curves may be used [29–31], or it may be assigned on the basis of the experience and exploitation history regarding the particular utility. The d_1 is an additional factor that depends on the general economic situation of the owner of the transformer. Its default value is 1, and if needed, it can adjust the relative importance of the economic aspect, depending on the specific market situation at a given time (the higher the d_1, the greater the influence of the economic aspect on the final EPI value).

The second part of (2) is related to the technical aspect of the analyzed transformer and uses d_2, n_i, and P_{tr} for calculation. P_{tr} represents the relative priority index of the assessed unit in the entire population. The default value of P_{tr} is 1—lower priority represents values below 1, and higher priority represents values greater than 1. The higher the value of P_{tr}, the greater the influence of the technical aspect on the final EPI rating. The importance of a specific input parameter from the transformer exploitation point of view is reflected in n_i. Its value theoretically can vary from 0 to 1, where 0 means "problem irrelevant to the exploitation perspective" and 1 means "decommissioning of the unit/unable to operate and repair". The d_2 is an additional factor that depends on the general technical situation of the owner of the transformer. For example, it can reflect the current availability of certain resources (spare parts, spare units, repair crews, the possibility of stoppage, etc.). Its default value is 1, and if needed, it can adjust the relative importance of the technical aspect depending on the specific situation at a given time (the higher the d_2, the greater the influence of the technical aspect on the final EPI value). Figure 2 illustrates the EPI calculation process for a given transformer.

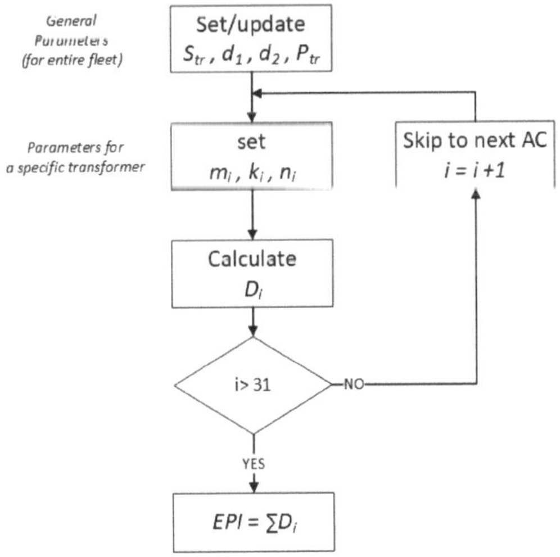

Figure 2. Flowchart of initialization of parameters for EPI calculation of the specific transformer.

Multiplication by 100 in (2) is not relevant from the EPI method point of view. However, in this case, it was included at the request of the utility this method was developed for to make the interpretation of the EPI somewhat similar to the "transformer wear percentage".

To illustrate the step-by-step calculation of D_i, moisture in the oil (D_{18}) will be used below as an example, as it is one of the most common defects in oil transformers. Starting with the first part of (2), d_1 is set to 1 (default value), as the general economic situation of the owner is normal. The estimated value of m_{18}/S_{tr} is about 0.02, assuming that oil refining is recommended (if replacement of the oil is recommended, then it will be about 0.04), and the new unit value is approximately 1 mln € (typical 31 MVA unit). k_{18} is set to 0.5, as the economical relative aspect of this defect is rather minor (typical procedure of oil treatment is needed, which is commonly available and does not require any additional costs), and the age of the transformers is assumed as the mean of the fleet. So, the final value of the first part of (2) is now 0.01. For the second part of (2), d_2 is set to 1 (default value), as the general technical situation of the owner of the transformer is normal. n_{18} is set to 0.05, as this is a typical defect that, if controlled and fixed, does not significantly affect the exploitation of the transformer. The priority of the transformer is set to the default value, $P_{tr} = 1$. So, the final value of the second part of (2) is now 0.05. Finally, adding two parts and multiplying by 100 gives the value of $D_{18} = 6$.

Table 2 presents a complete catalog of all possible input parameters for EPI and their proposed default values D_i assigned for the transformer population the EPI method was designed for. These values were calculated using (1) and (2), and specific values or parameters of (2) are mainly based on the experience of the authors, the suggestions and requirements of the owner of the fleet, and the specificity of the target user market. They also consider contemporary transformer exploitation guides, the expertise of the authors, and specialists from the utility in such a way as to reflect the importance of the indicated problems in the analyzed transformer and also take into account economic aspects, such as repair costs, carrying amounts, etc. [32,33]. As a result, proposed D_i values are not universal and should be adjusted to the specific fleet in which this system is to be used. However, the weights proposed in this study can be used as a starting point or reference in any case.

2.2.3. Absolute Rating Scale

The proposed EPI method allows for two types of transformer evaluation: absolute and relative. The absolute scale is based on a typical three-point rating and uses strictly assigned boundaries for each rating (Table 3). After determining the value of the EPI, a given transformer unit is classified into one of three groups, divided in terms of its current technical condition and operational measures necessary to implement, aimed at ensuring its further trouble-free operation [33]. The first group includes transformers whose technical condition does not raise any major objections or requires only minor operational procedures that do not really affect the safety of their further operation. In practice, these are transformers for which no defects and faults have been shown, or only these that do not pose a direct threat to further trouble-free operation of a given transformer have been shown. The first group will include units for which the EPI parameter is less than 10—this group has been labeled *Does not require significant operational/investment procedures* and may be marked as green (in line with the usual practice in this regard [13]).

Table 3. Absolute EPI rating scale.

Rating	EPI	Group
1	<10	Does not require significant operational/investment procedures
2	10–50	Required maintenance/investments
3	>50	Significant operational/investment measures required

The second group is labeled *Required maintenance/investments*, and may be marked as yellow. It includes the transformers in relatively good but not perfect technical condition. It means that in order to ensure their continued reliable operation, it is recommended (sometimes even necessary) to implement in the near future some specific measures necessary for the safety of this unit. Failure to implement the procedures may, in the near future, result in an emergency shutdown of the transformer or its damage. The values of the EPI parameter that result in assigning a given transformer to this group range from 10 to 50.

The last, third group are classified units in the worst technical condition, indicating a relatively high probability of failure or requiring immediate key operational measures, without which their further operation may be at risk or even impossible. This group is labeled *Significant operational/investment measures required* and may be marked as red.

2.2.4. Relative Rating Scale

Regardless of the absolute scale, EPI supports the relative evaluation scale of the assessed transformers in the context of the entire population ($EPI_{\%}$). The idea of the $EPI_{\%}$ is to indicate the technical condition of the assessed unit compared to the rest of the population. It requires gathering the EPI for all units in the population in one table. This table is used to calculate the percentile of the EPI for the entire population. As a result, the user obtains statistical information ($EPI_{\%}$), which shows what percentage of all units are in better and worse technical condition than the analyzed one. Such information complements the absolute EPI results, which may not be fully informed, especially regarding the perspective of the entire population. Moreover, the use of percentiles enables one to easily evaluate the conventional rating scale (absolute scale)—one only needs to select adequate percentile boundaries for each scale (i.e., in the case of a typical 5-grade rating, percentiles can be selected as follows: 0–20, 20–40, 40–60, 60–80, 80–100). In fact, this approach yields dynamic rating scales, as the boundaries are defined in a relative way (percentiles)—as a result, the absolute values of the boundaries will dynamically change as new data (new EPI) feeds the database. In practice, gathering the EPI of all transformers in the population may be very difficult (time consuming) or even impossible to achieve over a rational time horizon. Thus, alternatively, some representative sample of the population may be assigned with respect to the technical condition (and age) structure of the entire population (Figure 2).

3. Results and Discussion

This section shows the verification of the EPI method based on real-life scenarios in comparison with two other methods. The first method (HI1) was an industry standard transformer condition assessment method, which uses a table-based health index (scoring matrix). It means that diagnostic criteria are predefined and divided into several relevant categories and have the same values (boundaries) for the entire population. Each category has a 5-level rating scale (0–4 points), and the final rating is a simple sum of the ratings of all categories (Table 4). The higher the rating, the worse the technical condition of the transformer. In this study, seven categories were used for HI1: oil, solid insulation, windings, core, OLTC, bushings, and others (so the final HI1 value is between 0 and 28). Further details on the HI1 method can be found in [13].

Table 4. Scoring matrix for the HI1 method.

Rating	HI1	Technical Condition
0	0–4	As new condition. Minimal Signs of ageing or deterioration
1	5–10	Good condition. Reliable operation expected for a lengthy period
2	11–16	Acceptable condition with significant signs of aging or deterioration. Consider condition-based maintenance
3	17–22	Poor Condition. Repair or replacement should be considered within the short term
4	23–28	Very Poor condition. High likelihood of failure.

The second method (HI2) employed an algorithm that used advanced oil diagnostics. In particular, it can be read as an improved standard method that uses three main diagnostic categories: basic diagnostics, active parts, and aging processes. Also, HI2 uses a number of predefined weights, which must be applied to different input parameters prior to the final rating calculation. Finally, the higher the rating, the worse the technical condition of the transformer. The final HI2 value is between 0 and 100%, which is divided into three states describing the technical condition of the diagnosed transformer (Table 5). More details on HI2 are presented in [34].

Table 5. Scoring matrix for the HI2 method.

Rating	HI2	Technical Condition
0	0–27	Good
1	27–57	Average
2	57–100	Poor

To illustrate the performance of this method, a specific transformer population was used—this population is a property of the utility the EPI method was designed for. The population consists of over 1500 units; their rated powers vary between 10 and 80 MVA; and their typical rated voltages are 115/16.5 kV (sometimes there are units with secondary rated voltages of 22 kV, 6 kV, or three-winding units). It was not practically possible to analyze the technical condition (to perform routine tests) of all of the units in the accepted time frames, so a representative sample was used instead. The selected sample statistically reflects the technical condition, age structure, and working conditions of the entire population and consists of 300 units (Figure 3).

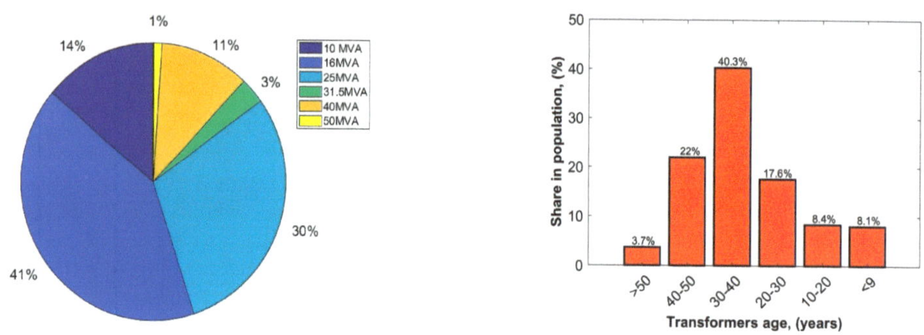

(**a**) (**b**)

Figure 3. Rated power (**a**) and age structure (**b**) of the representative transformer population.

For these units, a complete routine test was performed, the results of which were used to evaluate the EPI method. Figure 4 shows the normalized histograms of EPI results compared with the HI1 and HI2 methods. According to EPI, it is noticeable that two groups are dominant: units in very good technical condition (EPI < 20, which covers over 40% of the population) and units in quite bad technical condition (EPI > 60, which covers another approx. 40% of the population). The remaining units are in moderate technical condition (20 < EPI < 60). According to the presented method, the percentile value for a given EPI read from Figure 3 is called EPI%. According to HI1, almost 70% of the population is in relatively good condition (HI1 > 10). It is a typical situation with this kind of method. As it uses several categories, poor conditions related to only one category may be masked by relatively good results in other categories, and the final assessment may not be relevant in such a case. Comparing HI2 and EPI, the results are quite similar. However, in the case of HI2, it is also characteristic that most of the units were assessed as in good or average condition (HI2 < 57). This is probably the result of HI2 not taking into account the economic aspects of the transformer's technical condition and using only three diagnostic categories. On the other hand, HI1 and HI2, contrary to EPI, use raw measurement data to assess the transformer, which requires fixed diagnostic criteria. As a result, some specific malfunctions or defects may not be detected and included in the final assessment.

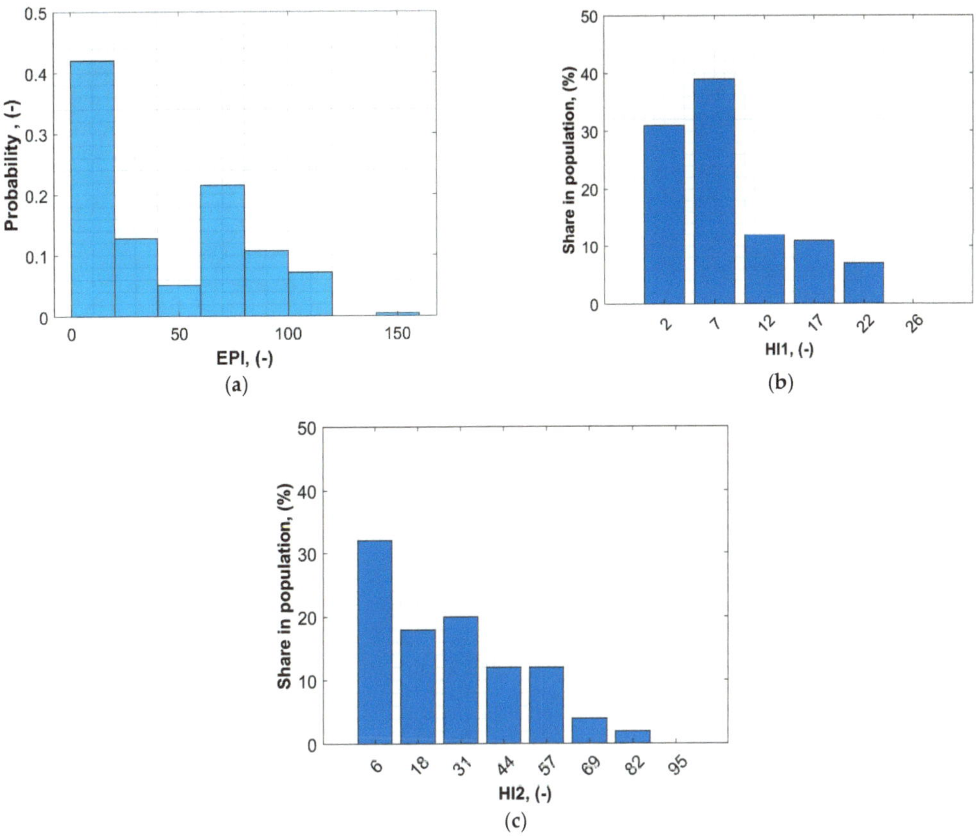

Figure 4. Histogram of technical condition assessment results in the analyzed population using different methods: (**a**) EPI, (**b**) HI1, and (**c**) HI2.

3.1. Use Case Tr1

This section describes an exemplary use case of the EPI method for the transformer in moderate conditions. In this case, the assessment was performed for a two-winding 40 MVA unit, 115/22 kV, which was manufactured in 2014 (Tr1). Table 6 presents an expert diagnosis and EPI results compared to HI1 and HI2 for Tr1. The diagnosis was based on the complete measurement data and visual inspection checklist issued after routine periodic tests. Expert diagnosis was used by the EPI to perform the assessment.

Table 6. Expert diagnosis and EPI results for Tr1.

Expert Diagnosis	i (According to Table 1)	D_i (According to Table 1)	EPI/EPI%	HI1	HI2
Moisture in the oil	18	6	26/45%	12	34
Moisture in solid insulation	24	20			

Tr1 was detected with two typical defects: moisture in the oil and in the solid insulation. These defects are quite common and usually do not imply any serious consequences, as long as they do not exceed specific thresholds. In this case, EPI is the sum of D_6 and D_{20} according to Table 1, which is 26. According to the proposed absolute EPI rating scale (Table 2), Tr1 has a rating of two, and is classified as *Required maintenance/investments*. To know the value of EPI% (in the proposed relative rating scale), a percentile distribution of the entire population must be known. In this case, the value of EPI% for Tr1 is 45%, which means that 45% of units in this population are in better technical condition than Tr1, and 55% are in worse condition (Figure 5). Results for HI1 and HI2 are quite similar to those for EPI. As the analyzed case is rather typical, both methods show that this transformer is in moderate condition and that some maintenance is needed in the near future.

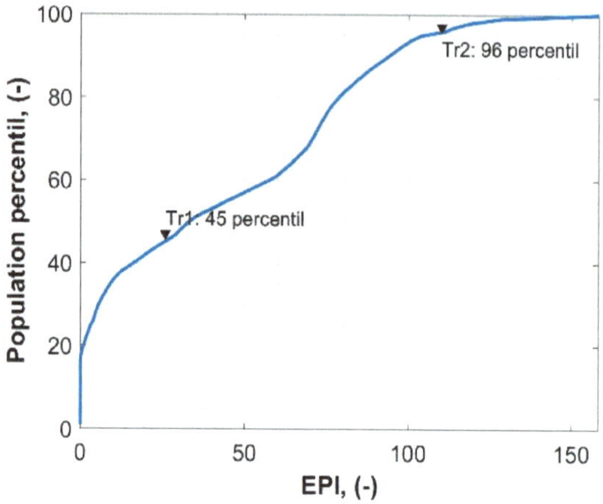

Figure 5. Percentile plot of EPI results in the analyzed population.

3.2. Use Case Tr2

This section presents the second exemplary use case of the EPI method for a transformer in poor technical condition. In this case, the EPI was performed for a three-winding 40 MVA unit, 115/33/16.5 kV, which was manufactured in 1976 (Tr2). Table 7 presents an expert diagnosis and EPI results for Tr2. As in the case of the analysis presented in Section 3.1, the diagnosis was based on the complete measurement data and visual inspection checklist issued after routine periodic tests. Additionally, in this case, the value of k_i was set to 1.1 considering the age of Tr2. These data were used by the EPI method to perform the assessment.

Tr2 was detected with 12 types of faults: minor oil leaks, minor and major paint loss or corrosion, cooling system malfunction, desiccant fault, oil leaks from the cooling system, moisture in the oil, aged cellulose, aged oil, moisture in the solid insulation, winding asymmetry, and OLTS defects. Some of these faults are trivial and do not affect the exploitation perspective of the transformer, but some might be read as serious and potentially imply the possibility of serious consequences for the safe exploration of the transformer. In this case, EPI resulted in 108 (considering $k_i = 1.1$). According to the proposed absolute EPI rating scale (Table 2), Tr2 has a rating of three and is classified as *Significant operational/investment measures required*. According to the relative rating scale, the $EPI_{\%}$ for Tr2 is $96_{\%}$, which means that 96% of units in this population are in better technical condition than Tr2, and only 4% are in worse condition (Figure 5). According to HI1, Tr2 was assessed as in poor condition. In this case, a very poor condition was expected, as this unit had a number of serious defects. Unfortunately, the sensitivity of HI1 is deficient in this case, and some serious issues are masked. As a result, HI1 may provide an optimistic indication of the transformer's condition. According to HI2, the score of Tr2 is 87, and it was the worst score in the analyzed population. On the other hand, EPI indicated that 4% of the population is in worse condition than Tr2. The economic aspect allowed the EPI to be more sensitive and accurate regarding this scenario.

Table 7. Expert diagnosis and EPI results for Tr2 (k_i = 1.1).

Expert Diagnosis	i (According to Table 1)	D_i (According to Table 1)	EPI/EPI% [1]	HI1	HI2
minor oil leaks (not within the main tank)	1	1	116/96% [1]	21	87
minor paint loss or corrosion	3	1			
major paint loss or corrosion	4	2			
cooling system malfunction	9	2			
desiccant faults (dehydrating breather)	11	0.1			
oil leaks (cooling system)	16	1			
moisture in the oil	18	6			
aged oil	19	7			
aged cellulose	23	60			
moisture in solid insulation	24	20			
windings asymmetry	28	3			
OLTC defects	30	2			

[1] considering k_i = 1.1.

4. Conclusions

This paper presents an alternative approach to the transformer assessment index. The EPI method was proposed and discussed. EPI is a relatively simple, dimensionless rating method for transformers. The proposed rating procedure is based on the results of actual periodical routine tests of the transformer, which makes it universal and easy to adopt for any transformer fleet. The method was designed for one of the European utilities where it has already been successfully implemented. As a result, its performance was verified in practice on 300 units of transformers. It was also compared with two different methods, and two representative use cases were described in this paper to confirm its usefulness. Regarding the presented study, some further conclusions may be drawn, as follows:

- EPI provides an absolute rating scale that corresponds with the current technical condition of the unit and its potential exploitation perspective;
- Simultaneously to the absolute rating scale EPI also provides a relative rating scale, which reflects the overall technical condition of the particular unit in the context of the entire fleet;
- Application of the EPI absolute rating scale requires gathering EPI for a representative sample of the population (ideally for all units in the population);
- EPI not only reflects the technical but also economic aspects of transformer maintenance;
- EPI can be potentially freely adopted for any transformer fleet, as well as for the specific situation of the utility, by adjusting the relevant parameters.

Apart from the advantages of the EPI mentioned above, some challenges and potential difficulties in application should be highlighted. Proposed D_i values are not universal and should be adjusted to the specific fleet in which this system is to be used, so implementation of EPI needs time, several optimization iterations, and expert knowledge. However, the weights proposed in this study can be used as a starting point or reference in any case. Furthermore, as EPI uses expert conclusions as input, it is also prone to any mistakes within the expert assessment. In the author's opinion, one promising solution would be the integration of the EPI with one of the decision support expert systems designed for power transformers.

Author Contributions: Conceptualization, M.K.; methodology, M.K.; software, M.K.; validation, M.K., S.B. and J.F.; formal analysis, M.K. and S.B.; investigation, M.K. and S.B.; resources, M.K. and S.B.; data curation, M.K. and S.B.; writing—original draft preparation, M.K.; writing—review and editing, M.K. and J.F.; visualization, M.K.; supervision, S.B.; project administration, S.B.; funding acquisition, M.K. and S.B. All authors have read and agreed to the published version of the manuscript.

Funding: This research received no external funding.

Institutional Review Board Statement: Not applicable.

Informed Consent Statement: Not applicable.

Data Availability Statement: Not applicable.

Conflicts of Interest: The authors declare no conflict of interest.

References

1. *WG A2.37*; Technical Brochure 642: Transformer Reliability Survey. CIGRE: Paris, France, 2015.
2. Freitag, S.C.; Sperandio, M.; Marchesan, T.B.; Carraro, R. Power transformer risk management: Predictive methodology based on reliability centered maintenance. In Proceedings of the SBSE 2018—7th Brazilian Symposium on Electrical Systems, Niteroi, Brazil, 12–16 May 2018; pp. 1–6. [CrossRef]
3. Murugan, R.; Ramasamy, R. Failure analysis of power transformer for effective maintenance planning in electric utilities. *Eng. Fail. Anal.* **2015**, *55*, 182–192. [CrossRef]
4. Piotrowski, T.; Rozga, P.; Kozak, R.; Szymanski, Z. Using the analysis of the gases dissolved in oil in diagnosis of transformer bushings with paper-oil insulation—A case study. *Energies* **2020**, *13*, 6713. [CrossRef]
5. Tenbohlen, S.; Coenen, S.; Djamali, M.; Müller, A.; Samimi, M.H.; Siegel, M. Diagnostic measurements for power transformers. *Energies* **2016**, *9*, 347. [CrossRef]
6. Samimi, M.H.; Ilkhechi, H.D. Survey of different sensors employed for the power transformer monitoring. *IET Sci. Meas. Technol.* **2020**, *14*, 1–8. [CrossRef]
7. Sylvestre N'cho, J.; Fofana, I.; Hadjadj, Y.; Beroual, A. Review of physicochemical-based diagnostic techniques for assessing insulation condition in aged transformers. *Energies* **2016**, *9*, 367. [CrossRef]
8. Kunicki, M.; Cichon, A. Analysis on partial discharges variability in mineral oil under long-term AC voltage. *IEEE Trans. Dielectr. Electr. Insul.* **2018**, *25*, 1837–1845. [CrossRef]
9. Kunicki, M.; Borucki, S.; Cichoń, A.; Frymus, J. Modeling of the Winding Hot-Spot Temperature in Power Transformers: Case Study of the Low-Loaded Fleet. *Energies* **2019**, *12*, 3561. [CrossRef]
10. Vantuch, T.; Fulneček, J.; Holuša, M.; Mišák, S.; Vaculík, J. An examination of thermal features' relevance in the task of battery-fault detection. *Appl. Sci.* **2018**, *8*, 182. [CrossRef]
11. Rozga, P.; Stuchala, F.; Piotrowski, T.; Beroual, A. Influence of Temperature on Lightning Performance of Mineral Oil. *Energies* **2022**, *15*, 1063. [CrossRef]
12. Fulnecek, J.; Misak, S. A Simple Method for Tree Fall Detection on Medium Voltage Overhead Lines with Covered Conductors. *IEEE Trans. Power Deliv.* **2021**, *36*, 1411–1417. [CrossRef]
13. *WG A2.49*; Techical Brochure 761: Condition Assessment of Power Transformers. CIGRE: Paris, France, 2019.
14. Prasojo, R.A.; Tamma, W.R.; Suwarno; Maulidevi, N.U.; Soedjarno, B.A. A method to calculate uncertainty due to unavailable data in transformer assessment index. In Proceedings of the International Symposium on Electrical Insulating Materials, Tokyo, Japan, 13–17 September 2020.
15. Prasojo, R.A.; Maulidevi, N.U.; Soedjarno, B.A.; Suwarno, S. Health Index Analysis of Power Transformer with Incomplete Paper Condition Data. In Proceedings of the 4th International Conference on Condition Assessment Techniques in Electrical Systems, CATCON 2019, Chennai, India, 21–23 November 2019. [CrossRef]
16. Azis Prasojo, R.; Suwarno; Ulfa Maulidevi, N.; Anggoro Soedjarno, B. A Multiple Expert Consensus Model for Transformer Assessment Index Weighting Factor Determination. In Proceedings of the 8th International Conference on Condition Monitoring and Diagnosis, CMD 2020, Phuket, Thailand, 25–28 October 2020. [CrossRef]
17. Aman, T.A.; Mokhtar, A.S.; Aizamtalib, M. Assessment the overall health condition of transformer using health index and critical index approach: TNB grid case study. In Proceedings of the PECon 2020—2020 IEEE International Conference on Power and Energy, Virtual, 7–8 December 2020. [CrossRef]
18. Tamma, W.R.; Prasojo, R.A.; Suwarno, S. Assessment of High Voltage Power Transformer Aging Condition Based on Health Index Value Considering Its Apparent and Actual Age. In Proceedings of the ICITEE 2020—12th International Conference on Information Technology and Electrical Engineering, Yogyakarta, Indonesia, 6–8 October 2020. [CrossRef]
19. Oliveira, M.M.; Bender, V.; Marchesan, T.B.; Kaminski, A.M.; Medeiros, L.H.; Wilhelm, H.M.; Neto, J.B.F. Power transformers assessment applying health index and apparent age methods. In Proceedings of the 2020 IEEE PES Transmission and Distribution Conference and Exhibition—Latin America, T and D LA 2020, Montevideo, Uruguay, 28 September–2 October 2020. [CrossRef]

20. Idrees, M.; Riaz, M.T.; Waleed, A.; Paracha, Z.J.; Raza, H.A.; Khan, M.A.; Hashmi, W.S. Fuzzy Logic Based Calculation and Analysis of Health Index for Power Transformer Installed in Grid Stations. In Proceedings of the RAEE 2019—International Symposium on Recent Advances in Electrical Engineering, Islamabad, Pakistan, 28–29 August 2019. [CrossRef]
21. Zeinoddini-Meymand, H.; Kamel, S.; Khan, B. An Efficient Approach with Application of Linear and Nonlinear Models for Evaluation of Power Transformer Health Index. *IEEE Access* **2021**, *9*, 150172–150186. [CrossRef]
22. Jian, W.; Wenbing, Z.; Chao, G.; Demeng, B.; Kuihua, W. The New Developed Health Index for Power Transformer Condition Assessment. In Proceedings of the 2020 5th Asia Conference on Power and Electrical Engineering, ACPEE 2020, Chengdu, China, 4–7 June 2020. [CrossRef]
23. Benhmed, K.; Mooman, A.; Younes, A.; Shaban, K.; El-Hag, A. Feature selection for effective health index diagnoses of power transformers. *IEEE Trans. Power Deliv.* **2018**, *33*, 3223–3226. [CrossRef]
24. Foros, J.; Istad, M. Health Index, Risk and Remaining Lifetime Estimation of Power Transformers. *IEEE Trans. Power Deliv.* **2020**, *35*, 2612–2620. [CrossRef]
25. Dong, M.; Zheng, H.; Zhang, Y.; Shi, K.; Yao, S.; Kou, X.; Ding, G.; Guo, L. A novel maintenance decision making model of power transformers based on reliability and economy assessment. *IEEE Access* **2019**, *7*, 28778–28790. [CrossRef]
26. Yahaya, M.S.; Azis, N.; Selva, A.M.; Kadir, M.Z.A.A.; Jasni, J.; Kadim, E.J.; Hairi, M.H.; Ghazali, Y.Z.Y. A maintenance cost study of transformers based on markov model utilizing frequency of transition approach. *Energies* **2018**, *11*, 2006. [CrossRef]
27. Kittan, S.; Kornhuber, S.; Kastel, P.; Nitsche, G.; Valtin, G.; Weise, M. Review and Implementation of Transformer Health Index Methods in line with the Development of a Condition Assessment Tool. In Proceedings of the 2018 International Conference on Diagnostics in Electrical Engineering, Diagnostika 2018, Pilsen, Czech Republic, 4–7 September 2018. [CrossRef]
28. *WG D1.39*; Technical Brochure 706: Guidelines for the Use of Statistics and Statistical Tools on Life Data. CIGRE: Paris, France, 2017; ISBN 9782858734085.
29. Lima, S.L.; Saavedra, O.R.; Miranda, V. A two-level framework to fault diagnosis and decision making for power transformers. *IEEE Trans. Power Deliv.* **2015**, *30*, 497–504. [CrossRef]
30. Taheri, A.A.; Abdali, A.; Rabiee, A. A Novel Model for Thermal Behavior Prediction of Oil-Immersed Distribution Transformers with Consideration of Solar Radiation. *IEEE Trans. Power Deliv.* **2019**, *34*, 1634–1646. [CrossRef]
31. Kornatka, M. Analysis of the exploitation failure rate in polish MV networks. *Eksploat. Niezawodn.–Maint. Reliab.* **2018**, *20*, 413–419. [CrossRef]
32. Borucki, S.; Cichoń, A.; Kunicki, M.; Frymus, J.; Skubis, J. Zastosowanie Narzędzi Sztucznej Inteligencji w Systemach Wspomagających Interpretacje Wyników Badań Diagnostycznych Transformatorów WN/SN. In Proceedings of the XII Konferencja Naukowo Techniczna Zarządzanie Eksploatacją Transformatorów, Wisła, Poland, 12–14 October 2022.
33. Frymus, J. Zintegrowany System Diagnostyki Sieciowej Wspomaganie Diagnostyki Transformatorów WN/SN. In Proceedings of the XII Konferencja Naukowo Techniczna Zarządzanie Eksploatacją Transformatorów, Wisła, Poland, 12–14 October 2022.
34. Bohatyrewicz, P.; Płowucha, J.; Subocz, J. Condition Assessment of Power Transformers Based on Health Index Value. *Appl. Sci.* **2019**, *9*, 4877. [CrossRef]

sensors

MDPI

Article

Voiceprint Fault Diagnosis of Converter Transformer under Load Influence Based on Multi-Strategy Improved Mel-Frequency Spectrum Coefficient and Temporal Convolutional Network

Hui Li *, Qi Yao and Xin Li

School of Electrical Engineering, Xi'an University of Technology, Xi'an 710048, China; 2211921149@stu.xaut.edu.cn (Q.Y.); 2211921145@stu.xaut.edu.cn (X.L.)
* Correspondence: lihui@mail.xaut.edu.cn; Tel.: +86-181-8262-6176

Abstract: In order to address the challenges of low recognition accuracy and the difficulty in effective diagnosis in traditional converter transformer voiceprint fault diagnosis, a novel method is proposed in this article. This approach takes account of the impact of load factors, utilizes a multi-strategy improved Mel-Frequency Spectrum Coefficient (MFCC) for voiceprint signal feature extraction, and combines it with a temporal convolutional network for fault diagnosis. Firstly, it improves the hunter–prey optimizer (HPO) as a parameter optimization algorithm and adopts IHPO combined with variational mode decomposition (VMD) to achieve denoising of voiceprint signals. Secondly, the preprocessed voiceprint signal is combined with Mel filters through the Stockwell transform. To adapt to the stationary characteristics of the voiceprint signal, the processed features undergo further mid-temporal processing, ultimately resulting in the implementation of a multi-strategy improved MFCC for voiceprint signal feature extraction. Simultaneously, load signal segmentation is introduced for the diagnostic intervals, forming a joint feature vector. Finally, by using the Mish activation function to improve the temporal convolutional network, the IHPO-ITCN is proposed to adaptively optimize the size of convolutional kernels and the number of hidden layers and construct a transformer fault diagnosis model. By constructing multiple sets of comparison tests through specific examples and comparing them with the traditional voiceprint diagnostic model, our results show that the model proposed in this paper has a fault recognition accuracy as high as 99%. The recognition accuracy was significantly improved and the training speed also shows superior performance, which can be effectively used in the field of multiple fault diagnosis of converter transformers.

Keywords: converter transformer; current; fault diagnosis; improved hunter–prey optimization; multi-strategy improved MFCC; voiceprint signal

Citation: Li, H.; Yao, Q.; Li, X. Voiceprint Fault Diagnosis of Converter Transformer under Load Influence Based on Multi-Strategy Improved Mel-Frequency Spectrum Coefficient and Temporal Convolutional Network. *Sensors* **2024**, *24*, 757. https://doi.org/10.3390/s24030757

Academic Editors: Pawel Rozga, Michał Kunicki and Jan Fulneček

Received: 26 December 2023
Revised: 16 January 2024
Accepted: 18 January 2024
Published: 24 January 2024

1. Introduction

In order to ensure the secure and reliable operation of converter transformers, which serve as intermediate devices for AC-DC power transmission technology, it is essential to conduct research on fault diagnosis. This research aims to enhance the accuracy and speed of fault identification, helping to promptly detect internal defects and prevent the further escalation of accidents. Unlike regular power transformers, converter transformers operate in a unique AC-DC working environment, which implies a higher level of harmonic currents. Consequently, this complexity in operational characteristics presents challenges in employing conventional fault diagnosis methods designed for traditional power transformers [1,2].

The converter transformer, in the process of operation with the core and windings, produces vibration because of electric power and other factors, and thus mechanical wave propagation through the transformer oil and rigid connection to the box. The resulting vibration and voiceprint signals contain a large amount of state information based on the vibration signal monitoring means that are widely used in the online monitoring of power equipment [3,4]. In the vibration signal acquisition process, the deployment location requirements of sensors are strict. Smaller deviations will interfere with the results. The noise detection method is used as a non-contact measurement; its sensor installation is convenient for solving the problem of high spatial sensitivity. At the same time, the voiceprint signal acquisition device has a wide frequency range to meet the monitoring requirements of different specifications of the transformer [5,6]. Numerous scholars both domestically and internationally have conducted research in this area, achieving promising results. In reference [7], four voiceprint emission feature spectra were constructed, and a lightweight fault diagnosis model was established to diagnose loose winding faults in transformers. Reference [8], based on the no-load operation of transformers, employed MFCC for voiceprint feature extraction, introduced Principal Components Analysis (PCA) to remove redundant features, and ultimately utilized the Vector Quantization (VQ) algorithm for accurate identification of loosened iron core faults. Reference [9] extracted features of on-load tap changers using Mel spectrograms and combined them with convolutional neural networks to recognize mechanical faults. However, the abovementioned voiceprint emission recognition techniques are based on traditional MFCC, which involves a cumbersome process of frame segmentation, windowing, and Fourier transformation to overcome spectral leakage issues. Furthermore, due to the inherent limitations of single-channel signal sources, the practicality of fault diagnosis using voiceprint emission signals is mostly limited to single-fault diagnosis.

To address the issue of the single-fault feature, reference [10] utilized Complete Ensemble Empirical Mode Decomposition (CEEMD) and short-time Fourier transform (STFT) to obtain temporal and spectral information about the signals. Deep fault features were then extracted using a deep fused convolutional neural network (DFCNN). Similarly, reference [11] proposed a mixed algorithm called high-order singular value decomposition (HOSVD)–high-order alternation least square (HOALS) to extract multi-dimensional features for pattern recognition. Furthermore, reference [12] combined the fusion multiscale convolutional neural network (F-MSCNN) to fuse sound and vibration features, leveraging the learning of multi-scale features for subsequent classification. Reference [13] proposed a real-time fault diagnostic method for hydraulic systems using data collected from multiple sensors in order to overcome the lack of information contained in a single sensor. Reference [14] processed signals from multiple sensors, thereby expanding the number of samples to enhance the diagnostic performance. However, most of the existing studies are based on single or homogeneous signals. They focus on extracting multidimensional features from different angles without considering multiple signal sources. The above diagnostic models do not start from different types of signal sources and ignore the correlation between different signals, making it difficult to extract deep information effectively from faults.

Existing approaches on data-driven fault classification mostly rely on artificial intelligence algorithms to analyze historical data and extract fault features, and the selection of parameters during the model training process has a crucial impact on the accuracy and convergence speed of fault classifiers. Reference [15] proposed a novel expectation maximization-unscented particle filter-Wilcoxon rank sum test (EM-UPF-W) method for data-driven techniques, which adaptively estimates noise variables with the help of the EM algorithm. References [16,17] used an artificial intelligence optimization algorithm for the adaptive optimization of machine learning parameters to avoid the human experience of parameter selection, but the existing artificial intelligence optimization is prone to the problem of local optimal stagnation, which has an impact on the final convergence speed and accuracy of the model.

Given this context, this article is focused on the division of current signals into intervals, combining voiceprint signals to achieve fault diagnosis in converter transformers. It overcomes the inherent limitations of single signal sources and conducts research on multi-fault diagnosis. The IHPO method is proposed to effectively address the local optimization problem, serving as a subsequent parameter optimization algorithm. VMD is employed for noise reduction, while the S-transform is utilized as a time-frequency conversion method. The improved MFCC technique based on multiple strategies is employed for feature extraction. ITCN is utilized for accurate fault identification, offering a novel approach for fault diagnosis in converter transformer systems. Furthermore, a specific 800 kV converter station was taken as a case study to validate the effectiveness of this integrated model.

The main contributions of this article are summarized as follows:

- This paper aims to counteract the problems of the traditional hunter–prey optimization algorithm, which easily falls into the local optimum, and of which the traversal of population initialization is not strong. It is improved via the introduction of SPM chaotic mapping and the Levy flight strategy, which is used for the adaptive selection of parameters in the fault diagnostic model to avoid the interference of the human experience selection.
- Multi-strategy improved MFCC is proposed for extracting voiceprint signals from converter transformers. Compared with the traditional voiceprint signal feature extraction method, the proposed approach incorporates the characteristics specific to the voiceprint signals of electric power equipment. It overcomes the interference of redundant information and demonstrates enhanced feature extraction capabilities.
- This paper introduces load signals to segment the operational intervals of converter transformers, realizes fault diagnosis through multiple types of signal sources, and proposes the improved multi-strategy MFCC and IHPO-VMD-ITCN fault diagnostic models. The experimental results demonstrate that the proposed fault diagnostic methods exhibit significant improvements in terms of both accuracy and calculation speed.

2. Analysis of Vibration Mechanism of Converter Transformer

Similarly to traditional power transformers, the vibration of converter transformers is induced by the electromagnetic forces in the windings and the expansion and contraction of the core due to magnetic hysteresis. These vibrations propagate through the transformer oil and rigid connections to the enclosure. However, owing to the complex environment resulting from the dual impact of alternating and direct currents, the vibration excitations are often characterized by multiple harmonic frequencies, leading to intricate vibration patterns in different areas.

2.1. Winding Vibration Mechanism Analysis

In accordance with the principles of high-voltage transmission, the current in converter transformers is accompanied by harmonic currents, including the $6k + 1(k = 1, 2, 3, \ldots)$th harmonic current at 50 Hz. This is manifested in Equation (1).

$$i = \sum I_{\alpha m} \cos(\alpha \omega_1 t + \Phi_\alpha) \quad (1)$$

where $I_{\alpha m}$ is the amplitude of each harmonic current, Φ_α is the phase angle of each harmonic, and ω_1 is the angular frequency of the 50 Hz current.

The interaction between currents of varying frequencies and magnetic fields generates axial and radial electromagnetic forces is expressed in Equation (2). The windings vibrate under the influence of these electromagnetic forces.

$$\begin{aligned} F_a &= k_{fa}(\sum I_{\alpha m} \cos(\alpha \omega_1 t + \Phi_\alpha))^2 \\ F_r &= k_{fr}(\sum I_{\alpha m} \cos(\alpha \omega_1 t + \Phi_\alpha))^2 \end{aligned} \quad (2)$$

where k_{fa} and k_{fr} represent the axial and radial electromagnetic force coefficient and F_a and F_r represent the winding axial and radial electromagnetic force.

Based on the motion differential equation, the acceleration of winding vibration can be represented by Equation (3):

$$\begin{aligned} a_a &= k_{aa}(\sum p_1 I_{\alpha m}{}^2 \cos(2\alpha\omega_1 t + \varphi_1) + \sum_2 p_2 I_{\alpha_1 m} \cos((\alpha_1 + \alpha_2)\omega_1 t + \varphi_2) \\ &+ \sum_2 p_3 I_{\alpha_1 m} I_{\alpha_2 m} \cos((\alpha_1 - \alpha_2)\omega_1 t + \varphi_3)) \\ a_r &= k_{ar}(\sum p_1 I_{\alpha m}{}^2 \cos(2\alpha\omega_1 t + \varphi_1) + \sum_2 p_2 I_{\alpha_1 m} \cos((\alpha_1 + \alpha_2)\omega_1 t + \varphi_2) \\ &+ \sum_2 p_3 I_{\alpha_1 m} I_{\alpha_2 m} \cos((\alpha_1 - \alpha_2)\omega_1 t + \varphi_3)) \end{aligned} \quad (3)$$

where $\sum_2$ is the sum of multiplication of different harmonics, k_{aa} and k_{ar} are the axial and radial acceleration coefficients, p_1, p_2, and p_3 are the calculation parameters, α_1, α_2 are the number of harmonics, and φ_1, φ_2, and φ_3 are the acceleration phase angles.

From Equation (3), it can be observed that under the influence of the $6k_1 + 1$ $(k_1 = 1,2,3,\ldots)$th harmonic, apart from the 100 Hz component, there is also a significant presence of the $100k_1$Hzth harmonic in the vibration of the converter transformer. When the natural frequency of the windings is close, resonance can easily occur, leading to a deviation of the dominant vibration frequency from 100 Hz.

2.2. Core Vibration Mechanism Analysis

The vibration of the core is primarily induced by magnetostriction. Furthermore, the excitation voltage of the converter transformer contains numerous harmonic components. Taking the influence of harmonic voltages into account, the vibration of the core can be represented by Equation (4):

$$\begin{aligned} a &= \frac{d^2(\triangle L)}{dt^2} = k_a(\sum 2U_{\alpha m}{}^2 \cos(2\alpha\omega_1 t + \varphi_\alpha) + \sum_2 q_1 \cos((\alpha_1 + \alpha_2)\omega_1 t + (\varphi_{\alpha 1} + \varphi_{\alpha 2})) \\ &+ \sum_2 q_1 \cos((\alpha_1 - \alpha_2)\omega_1 t + (\varphi_{\alpha 1} - \varphi_{\alpha 2})) \end{aligned} \quad (4)$$

Among them:

$$\begin{aligned} q_1 &= \frac{(\alpha_1+\alpha_2)^2 U_{\alpha_1 m} U_{\alpha_2 m}}{\alpha_1 \alpha_2} \\ q_2 &= \frac{(\alpha_1-\alpha_2)^2 U_{\alpha_1 m} U_{\alpha_2 m}}{\alpha_1 \alpha_2} \end{aligned} \quad (5)$$

where $U_{\alpha m}$ is the amplitude of each voltage harmonic, $\triangle L$ is the magnetostrictive deformation of the silicon steel sheet, and k_a is the saturation flux coefficient.

From Equation (4), it can be observed that the dominant frequency of the core vibration is primarily at 100 Hz. The influence of harmonics introduce a significant presence of the $100k_1$ Hz harmonic components. However, nonlinearities in the core and other factors may lead to deviations in vibration.

2.3. Fault Voiceprint Characterization of Converter Transformers

Similarly to ordinary power transformers, converter transformers are mainly composed of iron core, windings, and rigid connectors. When the iron core ages or experiences transportation and installation before operation, iron core loosening may occur. If the condition of iron core loosening is not promptly addressed, it will continue to accumulate, ultimately leading to iron core loosening failure. Iron core loosening failure results in a decrease in the fastening force between the silicon steel sheets of the iron core, thereby increasing the air gap between the stacked pieces. This causes a significant rise in the amplitude of iron core vibration acceleration, leading to changes in the intrinsic frequency of vibration and altering the voiceprint characteristics of the transformer. Similarly, during operation, the converter transformer is constantly subjected to the impact of electric power. In the event of a short-circuit fault, the intensification of electric power can prompt the occurrence of winding loosening faults. This leads to an aggravation of axial vibration, a significant increase in vibration acceleration amplitude, and changes in the vibration frequency distribution, resulting in alterations to the voiceprint characteristics of the transformer. When the converter transformer is running under bias magnetic conditions, the current signal can be regarded as the superposition of a DC component and Equation (1);

according to Sections 2.1 and 2.2 of the core and winding vibration mechanism analysis, it can be observed that, at this time, the vibration frequency of the converter transformer changes significantly.

In summary, when a fault occurs in the converter transformer, its core and winding vibration change significantly. The fault voiceprint signal generated under these conditions differs from that of normal operation. Therefore, the fault diagnosis of the converter transformer can be realized by adopting a machine learning algorithm for effective feature extraction of the voiceprint signal.

2.4. Characterization of Voiceprint Pattern Changes under Operating Conditions

The voiceprint signal and vibration signal, originating from the same source, exhibit a strong correlation. Based on the analysis in Sections 2.1 and 2.2, this study delves into the vibration characteristics of converter transformers during operation.

This study focuses on 28 converter transformers in a specific 800 kV converter station. Among them, there are 12 transformers per pole and 4 transformers on standby. The parameters of certain converter transformers are presented in Table 1.

Table 1. Parameters of converter transformer.

Parameters	Numerical Value	
	Pole II High-End Y/D Converter	**Pole I High-End Y/Y Converter**
Model number	ZZDFPZ-412300/600 kV	ZZDFPZ-412300/750/800
Rated capacity/MVA	412.3	412.3
Net side I_N/A	933	933
Valve side I_N/A	2357	4083
Operating frequency/Hz	50	50
Cooling method	OFAF	OFAF

The voiceprint signal acquisition system for the converter transformers is illustrated in Figure 1, and on-site acquisition photos are presented in Figure 2. We employed a combination of HS14401 capacitive sound sensors with a sampling frequency of 16 kHz along with a DHDAS dynamic signal acquisition instrument. Each converter transformer is equipped with three voiceprint acquisition devices, positioned on both sides and at a 45-degree angle, 0.5 m away from the enclosure. The data were collected in the outdoor substation environment under normal operating conditions, which may include noise interference. The voiceprint acquisition system was configured to collect voiceprint signals every 30 min, with each collection lasting for 60 s. Electrical parameters within the converter station were recorded every 30 min to ensure synchronization between the voiceprint signals and electrical parameters.

We selected time-length 0.1 s converter transformer in-operation voiceprint slices as the object of study. The time-domain and frequency-domain characteristics are illustrated in Figure 3. The main frequency of the converter transformer is 400 Hz, accompanied by a significant number of harmonics. This is attributed to the proximity of the winding intrinsic frequency to 400 Hz and the resonance of the converter transformer $100k_1$ Hz component, resulting in a deviation of 100 Hz compared to ordinary power transformers. This deviation corresponds to the theoretical analysis mentioned above.

The vibration characteristics of converter transformers vary under different operating conditions. In a no-load converter transformer, the core winding resonance becomes prominent. Under heavy load, the dominant vibration shifts to winding [18–20]. To facilitate a more precise quantitative analysis, this article focuses on the high-end Y/D converter transformer of pole II. The main objective is to analyze the main frequency change pattern of voiceprint characteristics concerning the magnitude of current. The results are depicted in Figure 4. Under no load, the main frequency of the converter transformer is 200 Hz, indicating the core vibration stage. At the rated voltage, when the valve side current is less than $0.2I_N$, the main frequency alternates between 200 Hz and 400 Hz. During this period, the core winding dominance alternates. However, when the current exceeds $0.23I_N$, the main frequency stabilizes at 400 Hz, signifying the dominance of winding vibration.

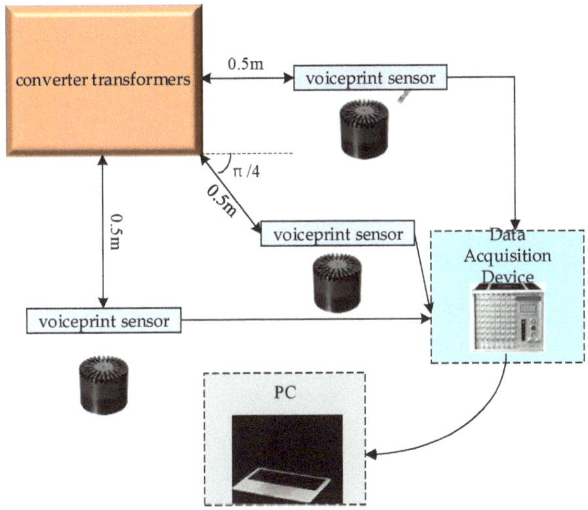

Figure 1. Voiceprint data acquisition system.

Figure 2. On-site acquisition.

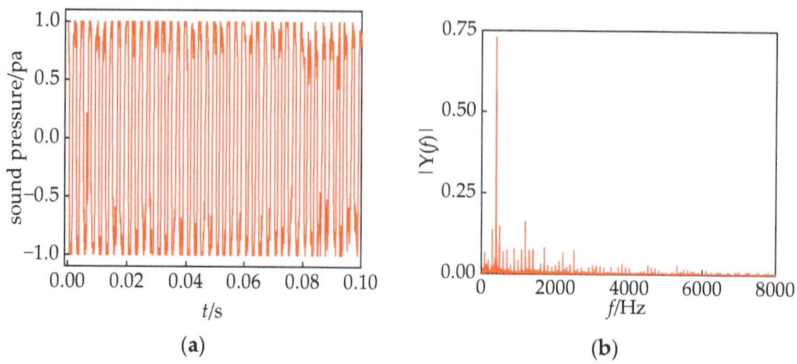

Figure 3. (**a**) Time domain characteristics; (**b**) frequency domain characteristics.

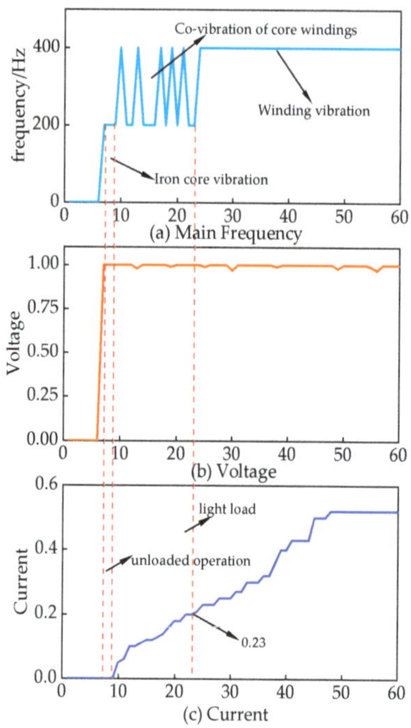

Figure 4. Characteristics of the main frequency of the voiceprint signal of the converter transformer with the variation in current.

Based on the information provided, a strong correlation exists between the electrical signals and voiceprint features of converter transformers. The division of converter transformers into three interval states, as illustrated in Table 2, allows for a phased approach to fault diagnosis. This approach proved effective in overcoming the issue of overlapping between core faults and winding faults, ultimately enhancing the accuracy of fault identification.

Table 2. Acousto-electric signal correlation analysis.

Operational State	No-Load (I)	Load (II)	Load (III)
Current and voltage signals	U = 1 I = 0	U = 1 I < 0.23	U = 1 I > 0.23
Voiceprint signal main frequency/Hz	200	200/400	400
Conclusion	Iron core vibration dominated	The core windings alternately dominate	Winding vibration dominant

3. Description of Fault Diagnosis Algorithms

3.1. Improved Hunter–Prey Optimization Algorithms

The hunter–prey optimization algorithm is a new intelligent optimization algorithm proposed by Naruei et al. in 2021 [21]. In this algorithm, the hunter adjusts its position to obtain the best hunting position, while the prey moves to a safe position to avoid the hunter's attack, and the safest position of the prey is the optimal solution of the problem to be optimized. This article proposes an improvement of the HPO algorithm by introducing the Levy flight strategy and SPM chaotic mapping. The modifications are briefly described as follows.

(1) Initialization: The conventional HPO algorithm achieves population initialization using Equation (6), as described below:

$$x_i = rand(1,d) \times (ub, lb) + lb \tag{6}$$

wherein x_i represents the positions of hunters or prey, d represents the problem dimensionality, and ub, lb represent the upper and lower bounds of the problem.

We chose Strongly Perturbed Mix (SPM) chaotic mapping for initializing the population, as shown in Figure 5. In comparison to circle mapping, the SPM demonstrates enhanced randomness and tergodicity, effectively addressing the issue of local clustering of individual hunters and prey [22]. The expression for SPM chaotic mapping is given by Equation (7).

$$x_{i+1} = \begin{cases} \mathrm{mod}(\frac{x(t)}{\eta}) + \mu\sin(\pi x(t) + r, 1), \\ 0 \leq x(t) \leq \eta \\ \mathrm{mod}(\frac{x(t)/\eta}{0.5-\eta}) + \mu\sin(\pi x(t) + r, 1), \\ \eta \leq x(t) \leq 0.5 \\ \mathrm{mod}(\frac{1-x(t)/\eta}{0.5-\eta}) + \mu\sin(\pi(1 - x(t)) + r, 1), \\ 0.5 \leq x(t) \leq 1-\eta \\ \mathrm{mod}(\frac{1-x(t)}{0.5}) + \mu\sin(\pi(1 - x(t)) + r, 1), \\ 1 - \eta \leq x(t) \leq 1 \end{cases} \tag{7}$$

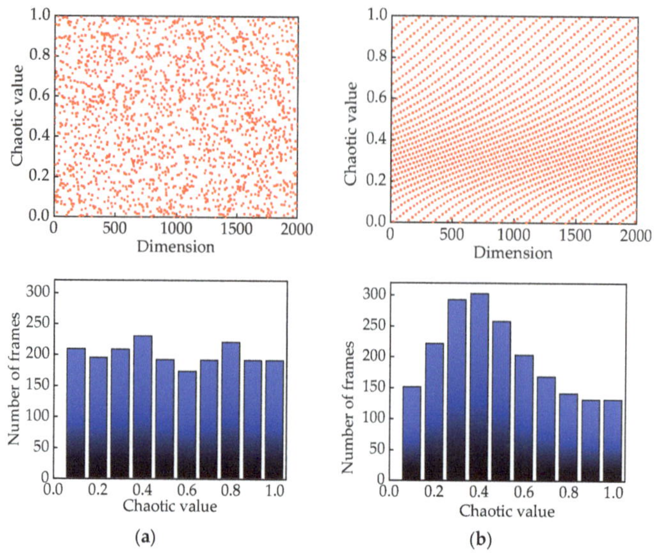

Figure 5. (**a**) SPM chaotic mapping value distribution; (**b**) circle chaotic mapping value distribution.

In Equation (7), the parameter $\eta \subseteq (0,1), \mu \subseteq (0,1)$ is typically chosen within the range of (0.4, 0.3).

(2) Optimization strategy: Hunters select prey that are far away from the group as their search targets, while the prey continuously move to evade hunter attacks and maximize their chances of survival. The position update for hunters and prey can be described by Equations (8) and (9), respectively.

$$x_{i,j}(t+1) = x_{i,j}(t) + 0.5[(2CZP_{pos(j)} - x_{i,j}(t)) + (2(1-C)Z\mu - x_{i,j}(t))] \tag{8}$$

wherein $x_{i,j}(t+1)$ represents the position of the ith hunter in the jth dimension at the (t + 1)th iteration, $x_{i,j}(t)$ represents the position of the ith hunter at the tth iteration, $P_{pos(j)}$ represents the position of the prey in the jth dimension, $C = 1 - 0.98t/T$ represents the balance parameter between exploration and exploitation, and Z is an adaptive parameter.

$$x_{i,j}(t+1) = T_{pos(j)} + CZ\cos(2\pi R_1) \cdot (T_{pos(j)} - x_{i,j}(t)) \tag{9}$$

wherein $T_{pos(j)}$ represents the global best position and R_1 represents a random number within the range of $[-1, 1]$.

It is challenging to overcome local optima solely by introducing SPM chaotic mapping. However, the utilization of the Levy flight strategy allows for a quick escape from local optima. The implementation approach is depicted in Equation (10).

$$Levy(s) \approx \frac{\lambda\beta(\Gamma(\lambda))\sin(\frac{\pi\lambda}{2})}{\pi} \cdot \frac{1}{s^{1+\lambda}} \tag{10}$$

wherein $\Gamma(\lambda) = \int_0^\infty t^{z-1}e^{-t}dt$ and the value of β is set to 1.5.

In practical applications, the Mantegna method is commonly used to generate random step lengths following a Levy distribution, as described in Equations (11) and (12).

$$S = \frac{\mu}{|v|^{\frac{1}{\beta}}} \tag{11}$$

$$\mu \sim N(0,\sigma^2), v \sim N(0,1)$$
$$\sigma = \left\{ \frac{\Gamma(1+\beta)\sin(\frac{\pi\beta}{2})}{\beta\Gamma(\frac{1+\beta}{2})2^{\frac{\beta-1}{2}}} \right\}^{\frac{1}{\beta}} \tag{12}$$

In the IHPO optimization algorithm, if the change in fitness values is continuously less than 0.001, the Levy flight strategy aids in escaping local optima. This generates the candidate solution for the next iteration, as shown in Equation (13).

$$x_i^{t+1} = x_i^t + \theta \oplus Levy(\beta) \tag{13}$$

In the equation, $\oplus$ denotes element-wise multiplication, θ is a random number uniformly distributed in the range [0, 1], and β is equal to 1.5.

The pseudocode used to improve the hunter–prey optimization algorithm is as follows in Algorithm 1:

Algorithm 1 Improve hunter–prey optimization

```
Input: HPO Parameters
Output: TargetScore, Best pos, Convergence curve
Initialize Hppos
Evaluate fitness of each HPpos
Set Target as the best HPpos, TargetScore as its fitness
for t = 2 to Max_iteration do
    Update c
    Update kbest
    for i = 1 to N do
    Generate random numbers
    if rand < B then
      Calculate xi and dist
      Set SI as HPpos(idxsortdist(kbest))
      Update HPpos(i,:) using formula with levy, l, c, z, SI, xi
        else
      for j = 1 to dim do
        Calculate v and rr
        Update HPpos(i,j) using formula with z(j), rr, Target(j), HPpos(i,j)
      end for
    end if
    Clip HPpos(i,:) values to be within bounds of lb and ub
    Evaluate fitness of HPpos(i,:)
    if HPposFitness(i) < TargetScore then
      Update Target and TargetScore
    end if
  end for
  Store TargetScore in Convergence curve(t)
end for
```

To validate the superiority of the IHPO algorithm, this article compares its performance with traditional optimization algorithms using the test function described in Equations (14) and (15). The results are depicted in Figure 6.

$$f_1(x) = -20\exp(-0.2\sqrt{\frac{1}{n}\sum_{i=1}^{n} x_i^2}) - \exp(\frac{1}{n}\sum_{i=1}^{n}\cos(2\pi x_i)) + 20 + e \tag{14}$$

$$f_2(x) = \frac{1}{4000}\sum_{i=1}^{n}(x_i^2) - \prod_{i=1}^{n}\cos(\frac{x_i}{\sqrt{i}}) + 1 \tag{15}$$

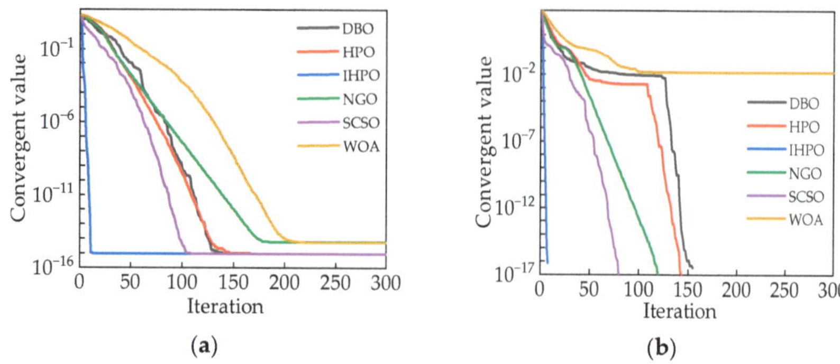

Figure 6. (**a**) Comparison of the optimization performance of the measurement function (14); (**b**) comparison of the optimization performance of the measurement function (15).

According to Figure 6a,b, it can be observed that the IHPO optimization algorithm converges to values of 8.9×10^{-16} and 0, respectively. The convergence speed of the IHPO algorithm is significantly higher than that of other traditional algorithms, achieving superior convergence values with the fewest number of iterations.

3.2. Variational Mode Decomposition

During the process of collecting transformed voiceprint signals, there is often a significant amount of noise interference. In order to ensure the accuracy of fault diagnosis, this article adopts the VMD algorithm for denoising processing, aiming to restore the original voiceprint signal as faithfully as possible.

The VMD algorithm constructs a variational problem and solves it [23,24]. Firstly, the original signal is decomposed into k modal components, denoted as $\mu_k(t)$. The energy spectrum is obtained through Hilbert transformation. $f(t)$ is made equal to each modal component $\mu_k(t)$ as a constraint condition, and the Lagrange multiplier $\lambda(t)$ and penalty factor α are introduced to transform it into a variational problem, as shown in Equation (16).

$$\begin{cases} \min\limits_{\{\mu_k\}\{\omega_k\}} \left\{ \sum\limits_{k}^{K} \left\| \partial_t [(\delta(t) + \frac{j}{\pi t}) * \mu_k(t)] e^{-j\omega_k t} \right\|_2^2 \right\} \\ \underset{\mu_k,\omega_k\lambda}{L} = \alpha \sum\limits_{k}^{K} \left\| \partial_t [(\delta(t) + \frac{j}{\pi t}) * \mu_k(t)] e^{-j\omega_k t} \right\|_2^2 \\ + \left\| f(t) - \sum\limits_{k=1}^{K} \mu_k(t) \right\|_2^2 + \left\langle \lambda(t), f(t) - \sum\limits_{k=1}^{K} \mu_k(t) \right\rangle \end{cases} \tag{16}$$

In Equation (16), * represents the convolution operation, $\mu_k(t)$ is the k-th modal component, ω_t is the central frequency, $\delta(t)$ is the impulse function, ∂_t represents the partial derivative with respect to t, and $\left\langle \lambda(t), f(t) - \sum\limits_{k=1}^{K} \mu_k(t) \right\rangle$ denotes the inner product.

The alternating direction multiplier method is used to solve the variational problem to find the optimal values of $\mu_k(t)$, ω_k, which is realized in the following steps.

(1) Initialize the parameters $\mu_k(t)$, ω_k, λ, set the loop $n = n + 1$, and iteratively update the parameters according to Equations (17)–(19).

(2) Update $\mu_k(t)$.

$$\hat{\mu}_k^{n+1}(\omega) = \frac{\hat{f}(\omega) - \sum\limits_{i=1}^{k-1} \hat{\mu}^n(\omega) + \frac{\hat{\lambda}^n(\omega)}{2}}{1 + 2\alpha(\omega - \omega_k^n)^2} \tag{17}$$

In Equation (17), $\hat{\mu}_k^{n+1}(\omega)$, $\hat{f}(\omega)$, $\hat{\lambda}^n(\omega)$ are the Fourier transforms corresponding to μ_k^{n+1}, $f(t)$, λ^n.

(3) Update ω_k.

$$\omega_k^{n+1} = \frac{\int_0^\infty \omega \left|\hat{\mu}_k^{n+1}(\omega)\right|^2 d\omega}{\int_0^\infty \left|\hat{\mu}_k^{n+1}(\omega)\right|^2 d\omega} \tag{18}$$

(4) Update λ.

$$\hat{\lambda}^{n+1} = \hat{\lambda}^n + \tau[\hat{f}(\omega) - \sum_{k=1}^{K} \hat{\mu}_k^{n+1}(\omega)] \tag{19}$$

(5) Determine convergence.

$$\frac{\sum_{k=1}^{K} \|\hat{\mu}_k^{n+1} - \hat{\mu}_k^n\|_2^2}{\|\hat{\mu}_k^{n+1}\|_2^2} < \varsigma \tag{20}$$

by setting $\varsigma > 0$.

(6) Determine whether the iteration condition is satisfied; if not, return to step (2).

3.3. *Multi-Strategy Improvement of MFCC for Dimensionality Reduction Extraction of Voiceprint Features*

As a common speech feature extraction method, MFCC is widely used in the field of speech recognition [25]. Considering that spectral leakage in the Fourier transform is very likely to occur, the S-transform is used as a time-frequency conversion method, and combined with the characteristics of the stationary energy of the converter voiceprint signal, it undergoes processing in the medium time to obtain the improved MFCC method to realize the voiceprint signal feature extraction.

3.3.1. S-Transform

The S-transform employs the Gaussian window function with adaptive adjustment of time and frequency parameters, replacing the fixed window function of the Fourier transform and the scale parameter window function of the wavelet transform. This approach exhibits higher-frequency characteristics at low frequencies and effectively improves the shortcomings of the Fourier transform [26].

The result of signal $x(t)$ after S-transformation is shown in Equation (21).

$$S(\tau, f) = \int_{-\infty}^{+\infty} x(\eta) w(\eta - \tau, f) e^{-j2\pi f \eta} d\eta \tag{21}$$

where f is the frequency, η is the time variable of $x(\eta)$, τ is the time component after S-transformation, and $w(\eta - \tau, f)$ is the Gaussian window function for adaptive adjustment, as shown in Equation (22):

$$w(\eta - \tau, f) = \frac{|f|}{\sqrt{2\pi}} e^{-\frac{(t-\eta)^2 f^2}{2}} \tag{22}$$

3.3.2. Multi-Strategy Improvement MFCC

In the field of audible sound recognition, given that the human ear exhibits varying sensitivities to the perception of each frequency band and the perception of the normal frequency band is nonlinear, Mel filtering is typically employed to transform the spectral information of voiceprint into Mel spectrum under Mel scale. The relationship between the normal frequency scale and the Mel frequency scale is expressed as in Equation (23):

$$Mel(k) = 2595 \times \lg(1 + f/700) \tag{23}$$

where f is the frequency on the regular scale and k is the frequency scale on the Mel scale.

In the domain of power equipment fault diagnosis, low-frequency information within 1000 Hz frequently incorporates numerous fault characteristics. Consequently, the utilization of Mel filters can adjust voiceprint information to varying degrees, enhance low-frequency information, and filter high-frequency information and compress it. The equal-height Mel filter bank function is expressed in Equation (24):

$$H_{(m)f} = \begin{cases} 0, f < x(m-1) \\ \frac{f-x(m-1)}{x(m)-x(m-1)}, x(m-1) \leq f \leq x(m) \\ \frac{x(m+1)-f}{x(m+1)-x(m)}, x(m) < f \leq x(m+1) \\ 0, f > x(m+1) \end{cases} \tag{24}$$

where m is the filter bank number and the number of filters in this paper is set to 26; therefore, the range of m is $0 < m < 26$, the center frequency of the Mel filter. The formula for the calculation of $x(m)$ is:

$$x(m) = (\frac{N}{f_s})Mel^{-1}(Mel(f_{\min}) + m\frac{Mel(f_{\max}) - Mel(f_{\min})}{\text{M+1}} \tag{25}$$

where f_s is the sampling frequency, $f_{\max}$, $f_{\min}$ represent the frequency range of the Mel filter bank, N is the number of S-transform samples, and M is the number of Mel filters.

The improved MFCC feature extraction method is distinguished from MFCC by the simpler operations of frame splitting and window adding. The specific steps are as follows:

(1) Framing: the S-transform has a high time complexity, so in order to save time, the original signal is framed with a fixed frame length.
(2) S-transform: the S-transform is performed on each frame by Equation (16) to obtain the time-frequency matrix $A(t, f)$.
(3) The spectral information is sought, as shown in Equation (26).

$$F(f) = \frac{\sum_{i=1}^{t} |A(t, f)|^2}{t} \tag{26}$$

where $A(t, f)$ is the time-frequency matrix, t is the time corresponding to the S-transform matrix, and f is the frequency.
(4) Bandpass filtering is performed, as in Equation (27).

$$Mel(m) = \ln(\sum_{k=0}^{N-1} |F(f)|H_m(f)) \tag{27}$$

where $Mel(m)$ is the Mel filter output and $H_m(f)$ is the filter bank.
(5) A discrete cosine transform is performed as in Equation (28) to obtain the first set of voiceprint characterization coefficients $feat_1$.

$$C(i) = \sum_{j=1}^{m} Mel(m) \cos(\frac{\pi i(m-0.5)}{26}) \tag{28}$$

(6) We perform first-order and second-order differentiation operations on $feat_1$ to obtain the second and third sets of parameters $feat_2$, $feat_3$ of the improved MFCC eigenvectors.
(7) We splice the three sets of parameters to form the feature vector IMFCC $= [feat_1, feat_2, feat_3]$.

Compared with the human speaking voice, power equipment voiceprint signal characteristics tend to be stationary; the feature vector obtained above contains a large amount of redundant information between the frames, so the use of mid-time features as shown in Equation (29) is more in line with the characteristics of stationary power equipment voiceprint features, reducing the interference of the heterogeneous long frames and having a stronger generalization [27], The multi-strategy improvement MFCC flowchart is shown in Figure 7.

$$MIMFCC = \frac{\sum_{i=1}^{N} IMFCC_i}{N} \tag{29}$$

where $IMFCC_i$ is the ith frame signal feature and N is the number of medium-time signal frames and denotes MIMFCC is the medium-time feature vector.

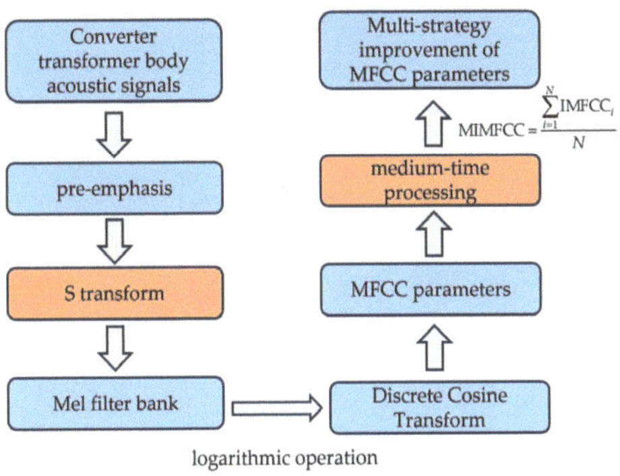

Figure 7. Multi-strategy improvement of MFCC flowchart.

3.4. Improved Temporal Convolutional Neural Networks

Time convolutional networks have good sequence information processing capabilities. In comparison to traditional architectures such as convolutional neural networks, this network achieves deeper networks by incorporating skip connections of residual blocks, effectively integrating shallow features into the depths for improved accuracy [28,29]. To simplify the network's complexity, cavity convolution is employed to expand the sensory field, and the causal cavity convolution is calculated as shown in Equation (30):

$$F(t) = \sum_{i=0}^{k-1} f(i) x_{t-di} \tag{30}$$

where d is the void coefficient, k is the convolution kernel size, and $f(i)$ is the ith element of the convolution kernel.

The traditional TCN residual module introduces nonlinearity through the Relu activation function. However, when the input is negative, the zero-gradient problem occurs, leading to the offset phenomenon. This, in turn, limits the learning efficiency and effectiveness of the TCN. Setting the output mean of the activation function to zero serves a dual purpose: it reduces the gradient vanishing problem and mitigates the impact of weight initialization. Additionally, the output of the activation function with zero-mean facilitates the propagation of information between the different layers of the network, resulting in better learning dynamics. This helps the network learn complex features and representations more efficiently. To a greater extent, it can enhance the network's learning performance. Therefore, the Mish activation function is used to replace the traditional Relu function, as in this equation:

$$F(x) = \mathrm{mish}(x) = x \times \tanh(\ln(1 + e^x)) \tag{31}$$

As depicted in Figure 8, compared with other activation functions, although the Tanh function has an absolute 0-mean value, it is prone to gradient vanishing due to the range of $[-1, 1]$. The Mish activation function is a better trade-off between the 0-mean value and the gradient vanishing problem [30].

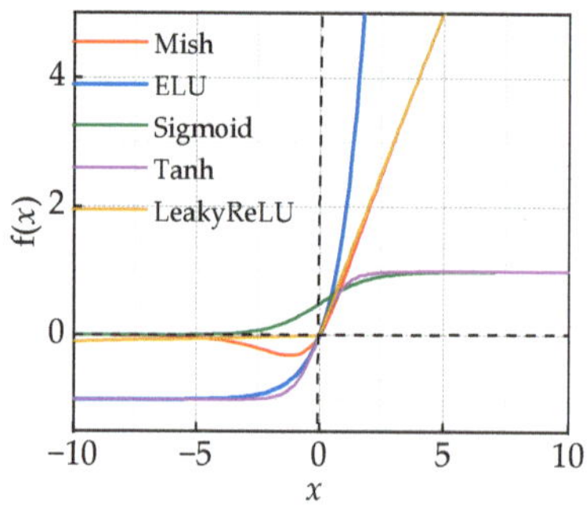

Figure 8. Activation function 0-mean comparison.

The improved TCN architecture is illustrated in Figure 9 ($k = 2$, $d = 1, 2, 4$), where each residual module contains two causal convolutional layers. The network's performance is enhanced through the incorporation of the Mish activation function, weight normalization, and dropout.

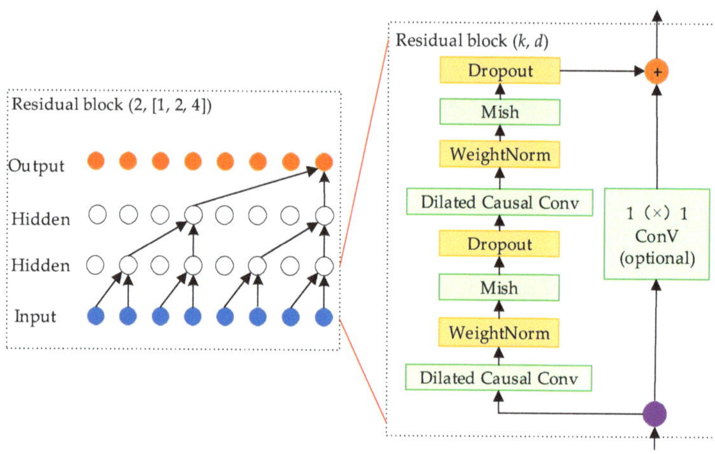

Figure 9. Improve temporal convolutional neural network architecture.

The improved TCN pseudocode is shown in Algorithm 2:

Algorithm 2 improved Temporal Convolutional Network

Input: Input sequence X with length T, Number of residual blocks K, Stack size S, Number of output channels C, Filter size f, Initial dilation value d_0, Learning rate η
Output: Probability distribution over classes
Initialize all model parameters
Set learning rate to η
Set initial dilation value to d_0
for $k = 1$ **to** K **do**
  for $s = 1$ **to** S **do**
    for $c = 1$ **to** C **do**
      Apply causal convolution to input sequence X with dilation d
      Apply activation function (e.g., Mish) to the output
      Apply weight normalization to the output
      Update output sequence O
    end for
  end for
  Stack the output sequence O with the input sequence X as the new input
  Increase the dilation value d exponentially
end for
Apply a fully connected layer to the final output sequence O
Apply softmax function to obtain probability distribution over classes

3.5. Multi-Strategy Improved MFCC-IHPO-VMD-ITCN Combined Fault Diagnosis Modeling

Converter transformer voiceprint signals are mainly concentrated in the low-frequency band. Considering the operating patterns of the converter transformer, a combined voiceprint–electric feature vector is adopted to overcome the problem of interference between core and winding vibrations. The accurate identification of converter transformer faults is achieved through a diagnostic process from denoising through feature extraction to pattern recognition. The diagnostic workflow is illustrated in Figure 10.

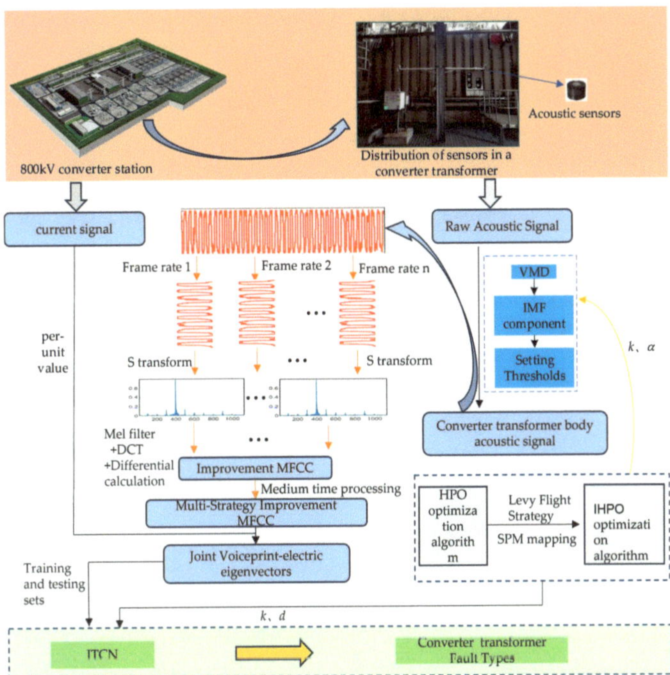

Figure 10. Multi-strategy improved MFCC-IHPO-VMD-ITCN fault diagnosis modeling.

The VMD is optimized based on IHPO to obtain the proprioceptive voiceprint signal. The selection of the decomposition number k and the penalty factor α has a significant impact on the decomposition result. It is prone to over-decomposition or loss of band information. Therefore, the minimum envelope entropy shown in Equation (32) is selected as the fitness function. IHPO is utilized to select the optimal $[k, \alpha]$ to overcome the inherent defects of VMD decomposition.

$$\begin{cases} Fitness = \min(f(i)) \\ f(i) = -\sum\limits_{i=1}^{N} p(i) \cdot \log 10(p(i)) \\ p(i) = a(i) / \sum\limits_{i=-1}^{N} a(i) \end{cases} \tag{32}$$

where N is the number of Intrinsic Mode Function (IMF) components, $f(i)$ is the envelope entropy after Hilbert adjustment, $p(i)$ is the normalized form, and $a(i)$ is the envelope signal.

Through the normalization of the load signal combined with the construction of multi-strategy improved MFCC for converter voiceprint and electric joint feature vector, multi-channel signal fault diagnosis is achieved.

Optimizing ITCN based on IHPO involves fine-tuning key parameters like kernel size (k) and dilation factor (d) for expansion convolution, which are crucial in determining the receptive field size and training accuracy. Utilizing Equation (33) as the fitness function enables adaptive optimization of ITCN to find optimal values for (k) and (d) that maximize the performance.

$$Fitness = (1 - accTrain) \times 100 \tag{33}$$

where $accTrain$ is the training set accuracy.

4. Calculus Analysis

4.1. Noise Reduction Processing for Voiceprint Signals

The voiceprint signals collected from outdoor substations are susceptible to significant transient and continuous noise interference, which inevitably affects the accuracy of fault diagnosis. Therefore, performing noise reduction processing is crucial.

Based on the given information, the optimization algorithm has a population size of 25 and a dimension of 2. The upper limit is denoted as $u_a = [25, 3000]$, while the lower limit is denoted as $u_b = [1, 500]$. Through 20 iterations, the fitness function changes are shown in Figure 11. In comparison to the HPO and HHO algorithms that converge to 3.208 and 3.2141, respectively, the proposed IHPO optimization algorithm in this study demonstrates better convergence performance.

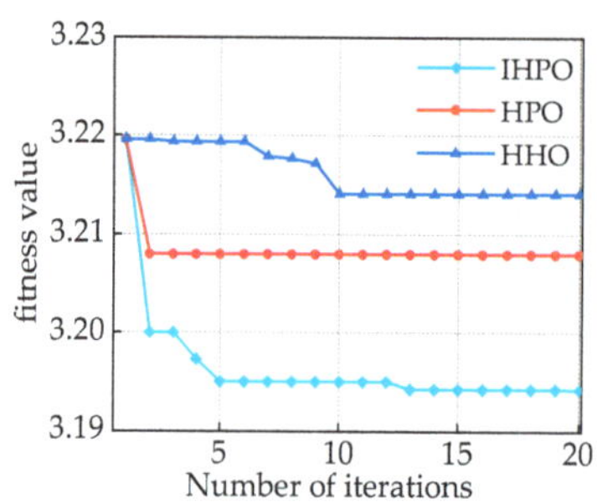

Figure 11. Comparison of fitness function values of different optimization algorithms.

It reaches the optimal solution within five iterations, with a significantly smaller final fitness value of 3.194. The optimal values obtained are k = 16 and α = 1246. The results of the IMF decomposition using IHPO-VMD are shown in Figure 12.

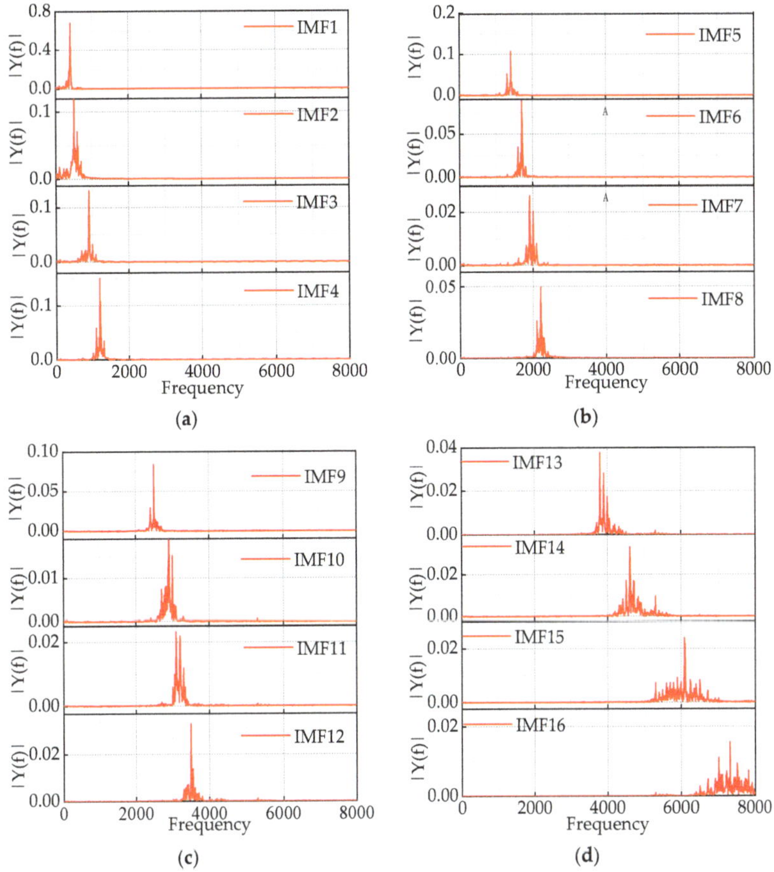

Figure 12. (**a–d**) Component IMF1–IMF16 after IHPO-VMD decomposition.

This article compares the results of IHPO-VMD with manually selected values of k and α to validate the superiority of IHPO-VMD. Taking $k = 16$ and $\alpha = 1000$ as an example, the first two decomposition results are shown in Figure 13a. When the value of α is too small, it results in a wide bandwidth, causing severe mode mixing between the 400 Hz and 500 Hz components, as well as between the 600 Hz and 1000 Hz components. In contrast, Figure 13b shows that IHPO-VMD avoids the mode mixing problem.

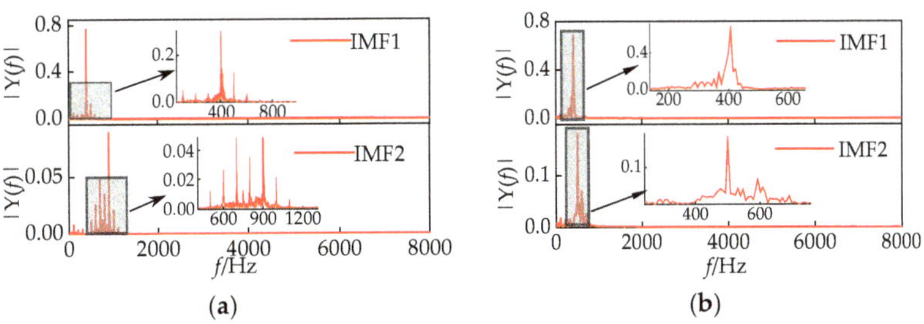

Figure 13. (**a**) Empirically selected VMD decomposition results; (**b**) IHPO-VMD decomposition results.

By calculating the correlation coefficients of the 16 IMF components, noise reduction processing can be achieved by setting a threshold using Equation (34). The correlation coefficients of each component are illustrated in Figure 14. Through the establishment of a threshold value, $C = 0.212$, the IMF1–IMF4 components can be recombined to derive the voiceprint signal of the converter transformer.

$$\begin{cases} C = \sqrt{\dfrac{\sum_{i=1}^{n} (\rho_i - \overline{\rho})^2}{k}} \\ \rho_k = \dfrac{\sum_{i=1}^{n} (x_{i,k} - \overline{x}_k)(y_i - \overline{y})}{\sqrt{\sum_{i=1}^{n} (x_{i,k} - \overline{x}_k)^2} \sqrt{\sum_{i=1}^{n} (y_i - \overline{y})^2}} \end{cases} \quad (34)$$

where ρ_i is the correlation coefficient of the ith order IMF component, $\overline{\rho}$ is the mean value, k is the number of components, x is the IMF component; y is the original signal; and n is the number of sampling points.

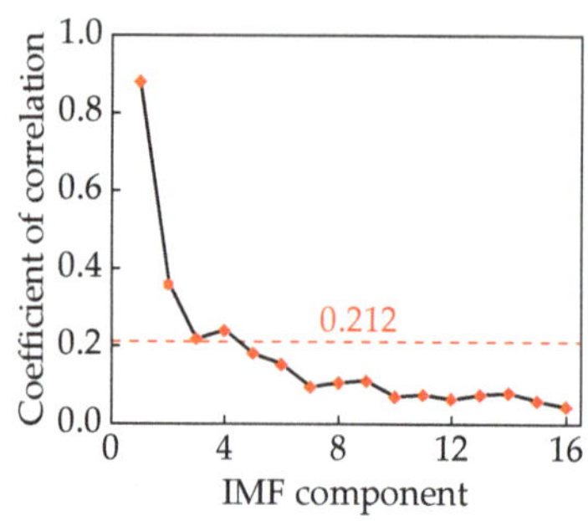

Figure 14. Correlation coefficients of components.

4.2. Joint Feature Vector Extraction

Based on the 1 s denoised voiceprint data, a frame length of 25 ms was chosen to generate an enhanced MFCC feature vector with a size of [36 × 39]. In this representation, 36 denotes the number of frames, and 39 signifies the dimensionality of the feature vector, as depicted in Figure 15a. The voiceprint signal of the converter transformer demonstrates stability, exhibiting high redundancy between frame numbers. To mitigate complexity, a mid-term feature vector of 250 ms was constructed, as depicted in Figure 15b, where the feature vector changes from [36 × 39] to [4 × 39]. This leads to a notable reduction in its complexity.

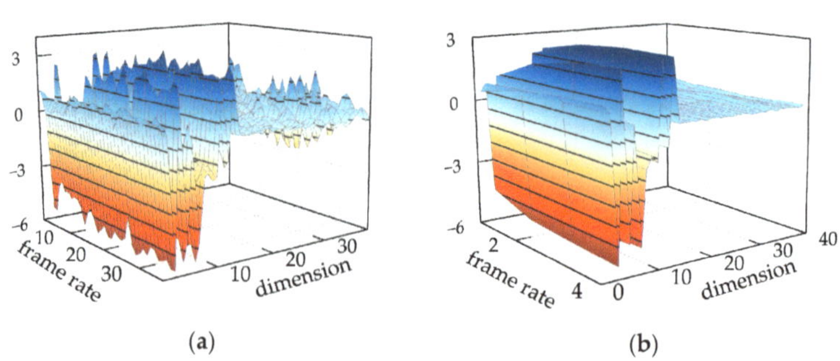

Figure 15. (**a**) Traditional MFCC features; (**b**) multi-strategy improved MFCC features.

Based on the provided information, feature extraction from the load signal was carried out using per-unit value to construct a joint voiceprint–electric feature vector of size $[4 \times 40]$. The iron core faults and winding faults in the converter transformer exhibit strong randomness, with distinct characteristic spectra corresponding to different loosening conditions. As analyzed in 2.3 and Table 1, iron core fault diagnosis is accomplished in Stage I, while winding fault diagnosis is achieved in Stage III. Through the separation of voiceprint features of the iron core and winding based on the load signal, a joint voiceprint–electric feature vector is constructed. In Stage II, the fault is defined as either an iron core or winding fault. However, this stage represents an unmonitorable phase, and determining whether a core failure or a winding failure is challenging for maintenance personnel. The fault diagnosis is conducted in stages to precisely identify iron core loosening faults, winding loosening faults, and DC bias faults. This approach effectively overcomes the limitation of existing research focusing on single fault diagnosis, providing a more comprehensive diagnostic capability. The spectral characteristics of typical defects in the converter transformer section are illustrated in Figure 16.

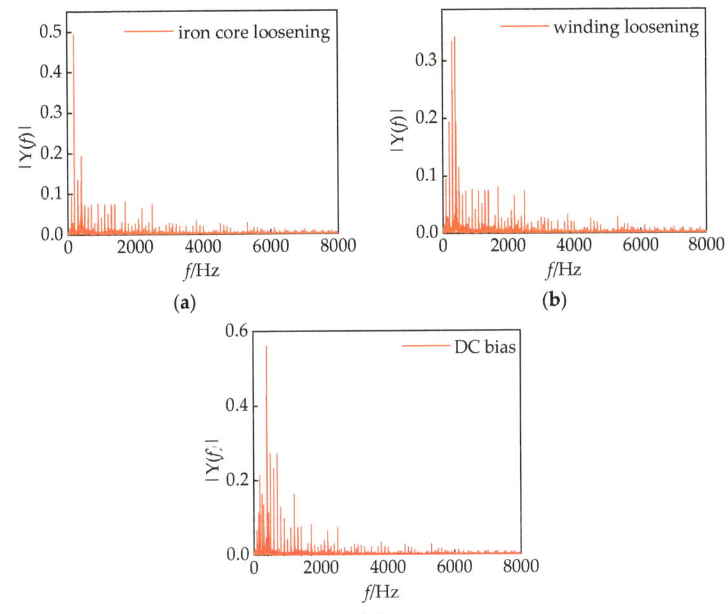

Figure 16. (**a**–**c**) are the typical defective spectral characteristics of the converter transformer.

4.3. Description of Experimental Objects and Measurement Points

The converter transformer, operating at a high voltage level and featuring a complex structure, plays a crucial role in high-voltage DC transmission technology. Utilizing the original model for fault diagnosis studies involves significant expenses and requires extensive equipment. Therefore, in this study, we sourced fault data from the signal detection system of an 800 kV converter station mentioned above. This system not only enables real-time storage of fault data but also allows for historical playback. To diversify fault samples, we used the monitoring system to collect fault signals from other converter stations to build a sample library. In this article, we collected fault signals from converter transformers experiencing DC bias, core loosening, winding loosening, and normal states. Both acoustic and current signals were collected through historical playback. The dataset was constructed following the method outlined in Section 4.2, involving division into the training sets and the test sets to ensure the effectiveness of deep learning [31], as shown in Table 3.

Table 3. Combined model training program.

Operational State	Serial Number	Training Sets/Each	Test Sets/Each
Normal	0	180	20
Iron core loosening	1	180	20
Winding loosening	2	180	20
DC bias	3	180	20
Core or winding fault	4	180	20

The IHPO parameters were set as follows: the number of populations is 30, the maximum number of iterations is 50, the epoch of parameter optimization is 50, the upper limit is $ua = [16, 6]$, and the lower limit is $ub = [1, 1]$. Adaptive optimization of the convolution kernel size k and the expansion factor d was realized, and the change in fitness value is shown in Figure 17, which converged to 0.082 after 21 iterations and gave outputs of $k = 16$ and $d = 3$.

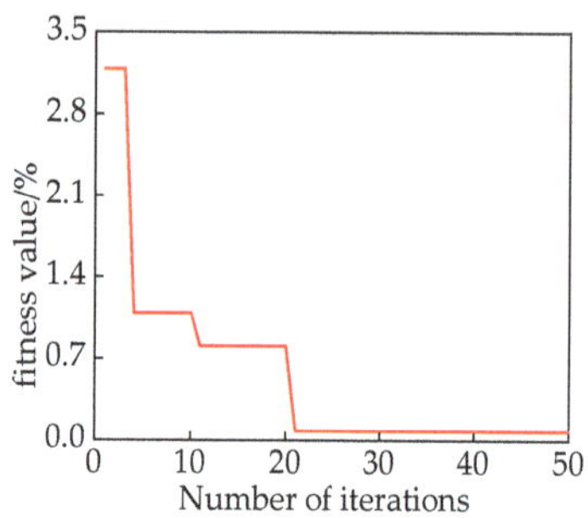

Figure 17. Change in fitness function.

The optimized results of IHPO were used as the input for ITCN, configuring the model with an epoch set to 100 and a batch size of 32. As illustrated in Figure 18a, this integrated model demonstrated stable convergence, achieving 100% accuracy after 88 epochs. To validate the training accuracy of the model, it was tested using a validation set, and the prediction results are depicted in Figure 18b, with a test accuracy of 99%. Through this analysis, the combination model, which utilizes current signals and incorporates audio–electric joint features, successfully mitigated interference between faults, affirming the feasibility of this combined model.

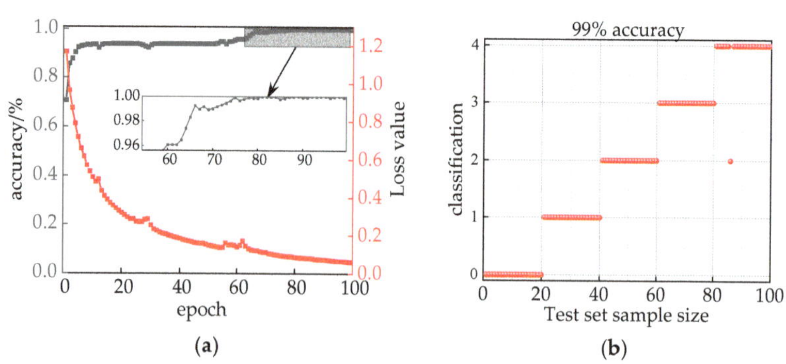

Figure 18. (**a**) Model identification results; (**b**) model testing set prediction results.

4.4. Comparative Analysis of Combined Forecasting Methods

To assess the performance of the combined model, in this article, we conducted a comparison with the IHPO-TCN model utilizing voiceprint–electric joint feature vectors and the IHPO-ITCN model based on audio feature vectors. The results are presented in Figure 19a. Upon comparing a (1) and a (2), it is evident that a (1) exhibits superior convergence, reaching 99.91% accuracy as epoch increases, surpassing a (2) in stability. This validates the superiority of model (a). In contrast, model a (3) achieves lower accuracy, converging to 95.41% after 94 epochs. The test set prediction results for the a (3) model are depicted in Figure 19b, with a test accuracy of 94%. Notably, mixed interference between core loosening and winding loosening faults is observed. In conclusion, the IHPO-ITCN model based on audio-electric joint feature vectors demonstrates significant superiority compared to the other models analyzed.

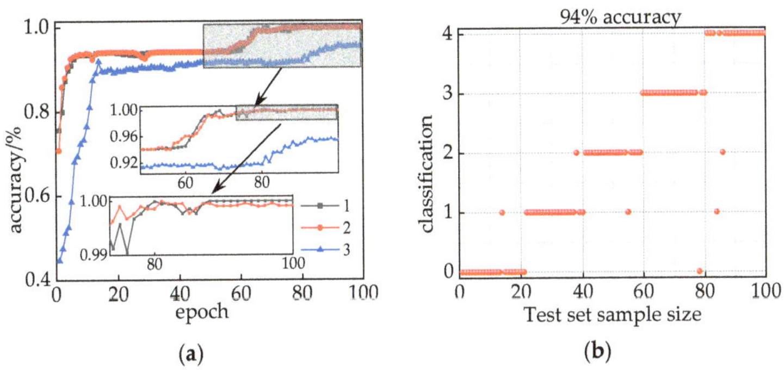

Figure 19. (**a**) 1 IHPO-ITCN based on joint voiceprint–electric feature vectors, (**a**) 2 IHPO-TCN based on joint voiceprint–electric feature vectors, (**a**) 3 IHPO-ITCN based on voiceprint feature vectors; (**b**) prediction results of the voiceprint feature model testing set.

A comparison of training time and accuracy of different feature signal fault recognition models is shown in Table 4. In the comparison experiments, the number of training sets and test sets are shown in Table 3, and the parameter settings of each model are also equal. Compared with traditional MFCC, MFCC's multi-dimensional improvement strategy decreased training time by 26 s and increased accuracy by 2.82%. These results validate the superiority of the improved MFCC in feature extraction. Due to changes in feature dimensions, the training time of feature vectors constructed by the voiceprint signals' combined load is longer. Compared with traditional MFCC features, traditional MFCC combined load features have a longer training time of 5.6 s but an accuracy improvement of 5.95%. Similarly, multi-strategy improvement MFCC combined load features have a training time increase of 1.1 s but an accuracy improvement of 4.33% compared to single multi-strategy improvement MFCC features. This verifies that although load signal intervention prolongs a certain training time, it effectively improves the accuracy of fault classification. For the diagnostic model proposed in this article, the accuracy ultimately converges to 100% and the training time is shorter, thus confirming the superior performance of the model.

Table 4. Comparison of training time and accuracy of different feature signal fault recognition models.

Characteristic Signal Type	Training Time/s	Convergence to Maximum Accuracy/%
Traditional MFCC	50.6	92.85
Multi-strategy improvement MFCC	24.6	95.67
Load + multi-strategy improvement MFCC	25.7	100
Load + traditional MFCC	56.3	98.8

In order to further substantiate the superiority of TCN in inverter voiceprint fault diagnosis, in this article, we conducted a comparative analysis with traditional machine learning algorithms, ensuring consistency in dataset determination, epochs, and other parameters used for the comparison method. The hyperparameter settings of the comparison model are provided in Table 5.

Table 5. Contrasting model hyperparameter settings.

Contrast Model	Activation Function	Batch Size	Learning Rate
TCN	Relu	16	0.001
CNN	Relu	16	0.001
LSTM	Relu	16	0.001
GRU	Relu	16	0.001

The recognition results of different machine learning models are presented in Table 6: Utilizing the load joint multi-strategy to improve MFCC parameters as fault features to construct a dataset, the four machine learning algorithms show good results in training time and test set recognition accuracy, further verifying the effectiveness of the fault diagnosis model in feature extraction. However, when compared with CNN, although TCN has a training time of 1.9 s longer, it excels in capturing deep features, leading to a 3% higher recognition accuracy. In contrast to TCN, the training times of GRU and LSTM are 2.2 s and 2.7 s longer, respectively, with accuracy reductions of 7% and 5%, confirming the superiority of TCN in this diagnostic model.

Table 6. Comparison model recognition results.

Contrast Model	Training Time/S	Test Set Accuracy/%
TCN	25.7	99
CNN	23.8	96
LSTM	27.9	92
GRU	28.4	94

5. Conclusions

This paper proposes a fault diagnosis method that combines the multidimensional-improvement strategy of MFCC with adaptive VMD-ITCN and incorporates the influence of load signals. This method significantly enhances recognition accuracy and is applicable in the field of fault diagnosis for converter transformers. Our experimental results demonstrate that the application of IHPO for optimizing VMD and ITCN has significant benefits, such as improved convergence and the avoidance of parameter-related impacts on fault diagnosis models. The introduction of load signals divides the entire operational process of the converter transformer into three stages, diagnosing core faults in Stage I and winding faults in Stage III. The effectiveness of the proposed model was verified using a sample dataset from an 800 kV converter station. This model exhibits superior performance in terms of recognition accuracy and training speed, providing a new approach for maintenance personnel to promptly and accurately detect internal defects in converter transformers.

The fault diagnosis model proposed in this article is based on a data-driven background, which achieves fault classification through row analysis of historical data of converter transformers. Therefore, the number of fault categories and samples is relatively small. In future research, we will collect fault data of converter transformers in different scenarios and expand the types of faults. The idea of transfer learning, as described in reference [32,33], can also be introduced to further improve the generalization of diagnostic models. On the other hand, we will consider establishing an accurate mathematical model from a model-driven perspective to simulate fault signals and achieve fault diagnosis.

Author Contributions: Data curation and writing—original draft preparation, Q.Y.; supervision and writing—review and editing, H.L.; validation, X.L. All authors have read and agreed to the published version of the manuscript.

Funding: This work is funded by the Ningxia Natural Science Foundation (No. 2022AAC03631).

Institutional Review Board Statement: Not applicable.

Informed Consent Statement: Not applicable.

Data Availability Statement: The data are contained within the article.

Acknowledgments: The authors are very grateful to the reviewers, associate editors, and editors for their valuable comments and time spent.

Conflicts of Interest: The authors declare no conflicts of interest.

References

1. Dai, H.; Li, N.; Han, W.; Song, Z.; Zhang, Y. Research on Fault Diagnosis and Analysis Method of Converter Transformer. In Proceedings of the 2021 IEEE International Conference on Electrical Engineering and Mechatronics Technology, ICEEMT 2021, Qingdao, China, 2–4 July 2021; pp. 314–317.
2. Sun, W.; Yang, L.; Zare, F.; Xia, Y.; Cheng, L. 3D modeling of an HVDC converter transformer and its application on the electrical field of windings subject to voltage harmonics. *Int. J. Electr. Power Energy Syst.* **2020**, *117*, 105581. [CrossRef]
3. Liu, X.; Sun, W.; Li, H.; Hussain, Z.; Liu, A. The Method of Rolling Bearing Fault Diagnosis Based on Multi-Domain Supervised Learning of Convolution Neural Network. *Energies* **2022**, *15*, 4614. [CrossRef]
4. Wang, C.; Cai, C.; Du, Y.; Ji, Z.; Liu, Y.; Miao, M. Fault Diagnosis of Transformer Winding Looseness Based on Vibration Signal and GOA-KELM Model. *J. Phys. Conf. Ser.* **2022**, *2290*, 012089. [CrossRef]
5. Zhang, P.; Jiang, W.; Zheng, Y.; Zhang, S.; Zhang, S.; Liu, S. Hydraulic-Pump Fault-Diagnosis Method Based on Mean Spectrogram Bar Graph of Voiceprint and ResNet-50 Model Transfer. *J. Mar. Sci. Eng.* **2023**, *11*, 1678. [CrossRef]
6. Jiang, W.; Li, Z.; Li, J.; Zhu, Y.; Zhang, P. Study on a Fault Identification Method of the Hydraulic Pump Based on a Combination of Voiceprint Characteristics and Extreme Learning Machine. *Processes* **2019**, *7*, 894. [CrossRef]
7. Ma, H.; Li, N.; Yang, Q.; Duan, D.; Zhu, H.; He, P. On-line fault diagnosis method of transformer winding looseness based on multi-characteristic voiceprint maps. *Electr. Mach. Control* **2023**, *27*, 76–87.
8. Wang, F.; Wang, S.; Chen, S.; Yuan, G.; Zhang, J. Voiceprint Recognition Model of Power Transformers Based on Improved MFCC and VQ. *Proc. Chin. Soc. Electr. Eng.* **2017**, *37*, 1535–1542.
9. Han, S.; Gao, F.; Wang, B.; Liu, Y.; Wang, K.; Wu, D.; Zhang, C. Audible Sound Identification of on Load Tap Changer Based on Mel Spectrum Filtering and CNN. *Power Syst. Technol.* **2021**, *45*, 3609–3617.
10. Chen, J.; Lin, C.; Yao, B.; Yang, L.; Ge, H. Intelligent fault diagnosis of rolling bearings with low-quality data: A feature significance and diversity learning method. *Reliab. Eng. Syst. Saf.* **2023**, *237*, 109343. [CrossRef]
11. Wang, W.; Li, Y.; Song, Y. Fault diagnosis method of vehicle engine via HOSVD–HOALS hybrid algorithm-based multi-dimensional feature extraction. *Appl. Soft Comput.* **2022**, *16*, 108293. [CrossRef]
12. Wan, H.; Gu, X.; Yang, S.; Fu, Y.A. Sound and Vibration Fusion Method for Fault Diagnosis of Rolling Bearings under Speed-Varying Conditions. *Sensors* **2023**, *23*, 3130. [CrossRef]
13. Tao, H.; Jia, P.; Wang, X.; Wang, L. Real-Time Fault Diagnosis for Hydraulic System Based on Multi-Sensor Convolutional Neural Network. *Sensors* **2024**, *24*, 353. [CrossRef] [PubMed]
14. Deng, C.; Deng, Z.; Lu, S.; He, M.; Miao, J.; Peng, Y. Fault Diagnosis Method for Imbalanced Data Based on Multi-Signal Fusion and Improved Deep Convolution Generative Adversarial Network. *Sensors* **2023**, *23*, 2542. [CrossRef] [PubMed]
15. Zhang, J.; Jiang, Y.; Li, X. Remaining Useful Life Prediction of Lithium-Ion Battery with Adaptive Noise Estimation and Capacity Regeneration Detection. *IEEE/ASME Trans. Mechatron.* **2023**, *28*, 632–643. [CrossRef]
16. Gao, S.; Xu, L.; Zhang, Y. Rolling bearing fault diagnosis based on SSA optimized self-adaptive DBN. *ISA Trans.* **2022**, *128*, 482–502. [CrossRef] [PubMed]
17. Song, Y.; Xie, H.; Zhu, Z. Predicting energy consumption of chiller plant using WOA-BiLSTM hybrid prediction model: A case study for a hospital building. *Energy Build.* **2023**, *300*, 113642. [CrossRef]
18. Wu, S.; Ji, S.; Sun, J.; Liang, N.; Zhao, T.; Dai, S. Vibration Monitoring and Variation Law of Converter Transformer in Operation. *High Volt. Eng.* **2022**, *48*, 1561–1570.
19. Jiang, P.; Zhang, Z.; Zhang, J.; Deng, B.; Pan, Z. Research on vibration characteristics and Multi-parameter state recognition of ± 500 kV converter transformer under fluctuating conditions. *Int. J. Electr. Power Energy Syst.* **2022**, *136*, 107748. [CrossRef]
20. Shi, Y.; Ji, S.; Zhang, F.; Li, J.; Han, S.; Ji, K. Multi-frequency Vibration Mechanism and Characteristics of Transformer Windings. *High Volt. Eng.* **2021**, *47*, 2536–2544.
21. Naruei, I.; Keynia, F.; Sabbagh Molahosseini, A. Hunter–prey optimization: Algorithm and applications. *Soft Comput.* **2022**, *26*, 1279–1314. [CrossRef]
22. Ban, D.; Lv, X.; Wang, X. Efficient Image Encryption Algorithm Based on 1D Chaotic Map. *Comput. Sci.* **2020**, *47*, 278–284.

23. Geng, G.; He, Y.; Zhang, J.; Qin, T.; Yang, B. Short-Term Power Load Forecasting Based on PSO-Optimized VMD-TCN-Attention Mechanism. *Energies* **2023**, *16*, 4616. [CrossRef]
24. Ouyang, M.; Shen, P. Prediction of Remaining Useful Life of Lithium Batteries Based on WOA-VMD and LSTM. *Energies* **2022**, *15*, 8918. [CrossRef]
25. Wan, S.; Dong, F.; Zhang, X.; Wu, W.; Li, J. Fault Voiceprint Signal Diagnosis Method of Power Transformer Based on Mixup Data Enhancement. *Sensors* **2023**, *23*, 3341. [CrossRef]
26. Minh Khoa, N.; Van Dai, L. Detection and Classification of Power Quality Disturbances in Power System Using Modified-Combination between the Stockwell Transform and Decision Tree Methods. *Energies* **2020**, *13*, 3623. [CrossRef]
27. Xu, M.; Li, Z.; Sun, H.; Sheng, G.; Jiang, X. Mechanical Fault Diagnosis Method of GIS Based on Improved MFCC. *High Volt. Appar.* **2020**, *56*, 122–128.
28. Lea, C.; Vidal, R.; Reiter, A.; Hager, G.D. Temporal convolutional networks: A unified approach to action segmentation. In *Computer Vision—ECCV 2016 Workshops, Proceedings of the ECCV 2016, Amsterdam, The Netherlands, 11–14 October 2016*; Lecture Notes in Computer Science (Including Subseries Lecture Notes in Artificial Intelligence and Lecture Notes in Bioinformatics); Springer: Cham, Switzerland, 2016; Volume 9915 LNCS, pp. 47–54.
29. Gao, Y.; Wang, W.; Lin, Q. Fault Diagnosis for Power Converters Based on Optimized Temporal Convolutional Network. *IEEE Trans. Instrum. Meas.* **2021**, *70*, 3501910.
30. Misra, D. Mish: A self regularized non-monotonic neural activation function. In Proceedings of the British Machine Vision Conference, Virtual, 7–10 September 2020; pp. 1–13.
31. Yu, Z.; Zhang, L.; Kim, J. The Performance Analysis of PSO-ResNet for the Fault Diagnosis of Vibration Signals Based on the Pipeline Robot. *Sensors* **2023**, *23*, 4289. [CrossRef]
32. Rezaeianjouybari, B.; Shang, Y. A novel deep multi-source domain adaptation framework for bearing fault diagnosis based on feature-level and task-specific distribution alignment. *Measurement* **2021**, *178*, 109359. [CrossRef]
33. Yang, B.; Lei, Y.; Li, X.; Li, N. Targeted transfer learning through distribution barycenter medium for intelligent fault diagnosis of machines with data decentralization. *Expert Syst. Appl.* **2024**, *244*, 122997. [CrossRef]

sensors

MDPI

Article

Measurement of Transient Overvoltages by Capacitive Electric Field Sensors

Felipe L. Probst *, Michael Beltle and Stefan Tenbohlen *

Institute of Power Transmission and High Voltage Technology (IEH), University of Stuttgart, 70569 Stuttgart, Germany; michael.beltle@ieh.uni-stuttgart.de
* Correspondence: felipe-luis.probst@ieh.uni-stuttgart.de (F.L.P.); stefan.tenbohlen@ieh.uni-stuttgart.de (S.T.)

Abstract: The accurate measurement and the investigation of electromagnetic transients are becoming more important, especially with the increasing integration of renewable energy sources into the power grid. These sources introduce new transient phenomena due to the extensive use of power electronics. To achieve this, the measurement devices must have a broadband response capable of measuring fast transients. This paper presents a capacitive electric field sensor-based measurement system to measure transient overvoltages in high-voltage substations. The concept and design of the measurement system are first presented. Then, the design and concept are validated using tests performed in a high-voltage laboratory. Afterwards, two different calibration techniques are discussed: the simplified method (SM) and the coupling capacitance compensation (CCC) method. Finally, three recorded transients are evaluated using the calibration methods. The investigation revealed that the SM tends to overestimate the maximum overvoltage, highlighting the CCC method as a more suitable approach for calibrating transient overvoltage measurements. This measurement system has been validated using various measurements and can be an efficient and flexible solution for the long-term monitoring of transient overvoltages in high-voltage substations.

Keywords: capacitive electric field sensor; capacitive divider; high-voltage measurement system; switching transients; transient overvoltages

Citation: Probst, F.L.; Beltle, M.; Tenbohlen, S. Measurement of Transient Overvoltages by Capacitive Electric Field Sensors. *Sensors* **2024**, *24*, 1357. https://doi.org/10.3390/s24051357

Academic Editors: Pawel Rozga, Michał Kunicki and Jan Fulneček

Received: 22 January 2024
Revised: 12 February 2024
Accepted: 13 February 2024
Published: 20 February 2024

1. Introduction

Transient overvoltage events are one of the major causes of dielectric failures in substation equipment [1,2]. These events are caused by switching operations, lightning discharges, and short circuits, among others [3]. With the increasing integration of renewable energy sources into the electric power system, the system dynamics are changing, and new transient phenomena are occurring [4–6]. Investigating their effects on substation equipment and accurately measuring the transients becomes even more important.

The required frequency bandwidth of the measuring device depends on the transient being measured. In high-voltage substations, voltage measurement is often carried out using capacitive voltage transformers (CVT) or inductive voltage transformers (IVT). Although these devices provide accurate measurements at nominal frequency, their outputs deviate significantly at higher frequencies [7–11]. Consequently, this characteristic restricts their application for measuring fast transients.

To overcome the limitations of voltage transformers, specific techniques have been proposed in the literature [10–16]. Most of these methods focus on correcting the distorted output voltage. However, these approaches often require complex filtering techniques or the installation of additional devices. Another strategy to measure transients is to use voltage sensors connected to the measurement tap of transformer bushings [17–21]. While this technique has high accuracy and wide bandwidth, its use is limited to transformer bays.

Using electric field sensors for voltage measurement is another solution that has been the subject of research and further improvement [22–33]. In [25–28], the authors propose voltage calculation through inversion or integration of the measured electric field. However, these methods require complex algorithms to solve matrix inverse operations or numerical integration of the spatial electric field. This results in low accuracy and long computation time [25,28]. In [29], the voltage is calculated from the rate of change in the electric displacement field, using a differential self-integration D-dot sensor. However, this sensor is a near-field device positioned on the ceramic casing of the conductor, exhibiting inflexibility and a potential galvanic risk. Another investigated technique involves voltage measurement based on a differentiating/integrating (D/I) concept [30–32], utilizing a capacitive pick-up electrode composed of a parallel plate capacitor with a grounded bottom plate. The primary drawback of this method is its low accuracy due to the neglect or inaccurate estimation of coupling between the outer phases. In [33], various alternative methods for measuring transients using unconventional transducers are discussed. One of these methods employs a coupling plane as an electric field sensor; however, the author provides only a brief overview of the technique.

This paper presents a measurement system based on capacitive electric field sensors to measure transient overvoltages in a high-voltage substation. Two prior publications [34,35] have briefly outlined the system's concept, major components, and its field application. In [34], the measurement system monitored switching transient overvoltages during the energization of a 420 kV transmission line. The primary aim was to investigate the failure of a surge arrester during a switching operation. In [35], the authors evaluated some measured transient overvoltages in the frequency domain. The analysis revealed that the transients contain significant high-frequency components (>>1 kHz), making traditional IVT or CVT unsuitable for accurately measuring them.

This study extensively discusses the concept and design of the measurement system, introduces an improved calibration method, and provides a statistical analysis of signals measured during a one-year measurement campaign. Section 2 outlines the system's concept and design. Section 3 reports the results of the validation performed in a high-voltage laboratory. Section 4 describes the aspects of the measurement system installation in a high-voltage substation and discusses two calibration methods. Section 5 presents the statistical analysis of all recorded transients and evaluates three of them using the different calibration methods. Finally, Section 6 presents the conclusions of the study.

2. Concept and Design of the Measurement System

A measurement system based on capacitive electric field sensors has been developed to measure transient overvoltages in substations. Figure 1 shows its schematic diagram.

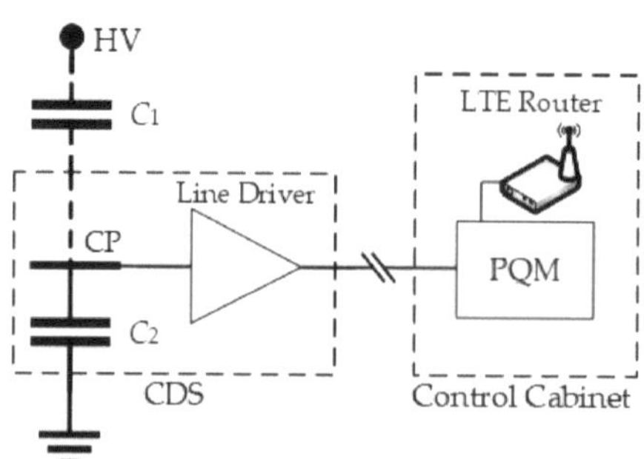

Figure 1. Schematic diagram of the transient overvoltage measurement system.

The capacitance C_1 represents the stray capacitance between the coupling plane (CP) and the high-voltage connection (HV) if the measurement device is installed under a substation busbar or on the base of a disconnector (see Figure 2). The user-defined capacitance C_2 is connected between CP and ground and is chosen based on the desired voltage ratio. There is also a stray capacitance between the CP and ground (around 30 pF), but it is much smaller than C_2 and can be neglected. Therefore, considering both capacitances C_1 and C_2, the measurement system has a capacitive divider whose voltage ratio is given by:

$$\frac{V_{in}}{V_{out}} = \frac{C_1 + C_2}{C_1} \quad (1)$$

where V_{in} is the voltage between HV and ground, and V_{out} is the divider's output voltage.

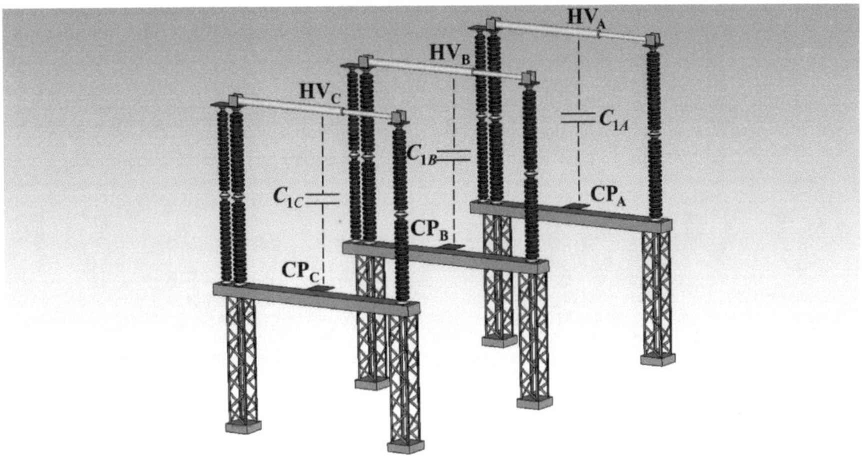

Figure 2. Three-dimensional model of the measurement setup for calculation of stray capacitance C_1.

Equation (1) is used to calculate the expected voltage ratio, considering the maximum voltage expected in a transient overvoltage ($V_{in\text{-}max}$) and the maximum output voltage suitable for the measurement system ($V_{out\text{-}max}$). In this study, $V_{in\text{-}max}$ is considered as 1 MV and $V_{out\text{-}max}$ as 20 V. Thus, the theoretical voltage ratio is 50,000:1. To accurately determine this voltage ratio, the capacitances C_1 and C_2 must be known. However, C_1 is a stray capacitance that changes according to the layout where the measurement system is installed. Therefore, this capacitance is estimated using the finite element method (FEM) in CST Studio Suite® version 2021 [36]. The three-dimensional model of the setup is shown in Figure 2.

The three sensors are installed, one on each base of a three-phase disconnector. The electric field can be assumed to be quasi-static considering the steady-state voltage. As a result, it exhibits the characteristics of an electrostatic field [25]. Therefore, the stray capacitance C_1 is calculated using the electrostatic solver in CST Studio.

The CST Studio model uses the technical drawing of the substation as the reference. The distance between phases is 4.50 m. The disconnector base is made of metal and has a height of 2.75 m. The moving contact of the disconnector is 3.75 m long and has a radius of 0.05 m. The insulators are modeled as ceramic porcelain with a relative permittivity $\varepsilon = 6$, a relative permeability $\mu = 1$, a height of 3.35 m, and an outer radius of 0.13 m. The sensor's CP is 0.10 m above the base of the disconnector. It has a length, width, and height of 0.40 m, 0.40 m, and 0.04 m, respectively. All the metal parts are modeled as perfect electric conductors (PEC), with the voltage of the moving contact set at 242.5 kV. The sensors and bases are grounded.

According to the electrostatic simulation, the calculated capacitances, C_{1B}, and C_{1C} are 0.766 pF, 0.772 pF, and 0.776 pF, respectively. The difference between C_{1A} and C_{1C} arises from the inherent approximation within the FEM. To achieve a voltage ratio of 50,000:1, the capacitance C_2 should be 38.6 nF, based on the average of the calculated C_1 values.

Additionally, there is a coupling between each sensor and the adjacent phases. This topic will be discussed in Section 4.

2.1. The Capacitive Divider Setup

After estimating the capacitance C_1 and defining the voltage divider requirements, the capacitive divider setup (CDS) is designed and assembled. The CDS comprises a CP and an electronic circuit mounted in an IP65 metal enclosure. The CP is an aluminum plate connected to the electronic circuit through an external terminal. Figure 3 shows the CP (a) and the metal housing (b) of the electronic circuit. The housing has four external connectors: one 4 mm connector for the CP, one N-connector for the measurement coaxial cable, one waterproof 2-pin connector for the DC source, and one ground connector.

(**a**)

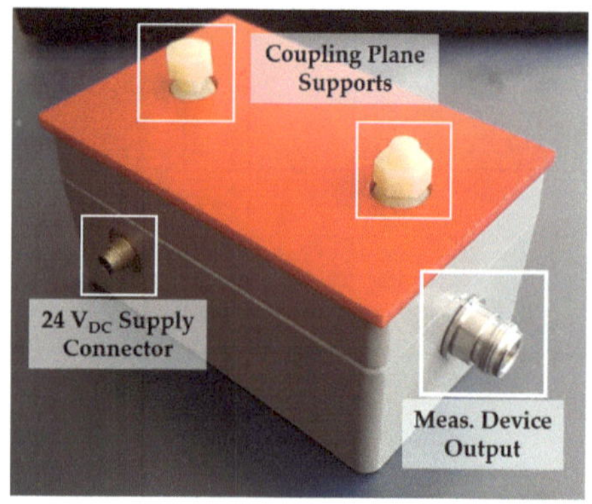

(**b**)

Figure 3. External components of the CDS. (**a**) Coupling plane; (**b**) metal housing with connections.

Inside the metal housing, the CP connects to the capacitors constituting C_2. Specifically, six 8.3 nF temperature-stable SMD capacitors are connected in parallel to achieve a capacitance of 49.8 nF. If the output of the voltage divider is directly connected to a measurement instrument using a long coaxial cable, the cable's capacitance influences the voltage ratio of the measurement system. Consequently, a line driver is employed to decouple the capacitive divider from the coaxial measurement cable.

The line driver uses a JFET input operational amplifier [37] with a slew rate of 60 V/µs, a bandwidth of 14 MHz, and a supply voltage of ±20 V. The circuit board also has additional internal buffered converters [38] to ensure voltage stability. A 15 m coaxial cable connects the line driver output to a power quality monitor. Figure 4 shows the electronic circuit of the CDS (a) and the six parallel capacitors of C_2 (b).

2.2. The Power Quality Monitor

A power quality monitor [39] is installed in a control cabinet to read and record the measured transient signals. It measures high-frequency harmonics and transients at a sampling rate of 1 MS/s. It has eight channels with synchronous sampling: four channels with 4 mm connectors with an input impedance of 10 MΩ and an input range from −600 V to +600 V; and four channels with BNC connectors with an input impedance of 1 MΩ and an input range from −50 V to +50 V. The transients recorded by the power quality monitor PQM-800 are stored internally and can also be uploaded to a monitoring cloud.

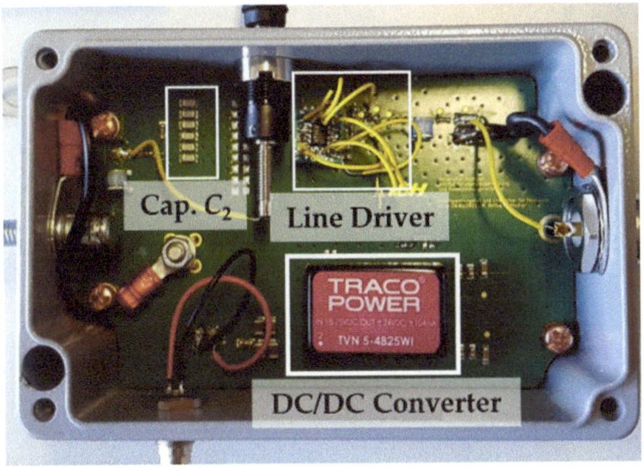

(a)

(b)

Figure 4. Electronic circuit of the CDS. (**a**) Top view; (**b**) parallel capacitors of C_2.

The control cabinet also includes an LTE router for sending the recorded transients to the monitoring cloud and a 24 V DC power supply for the CDS, as shown in Figure 5.

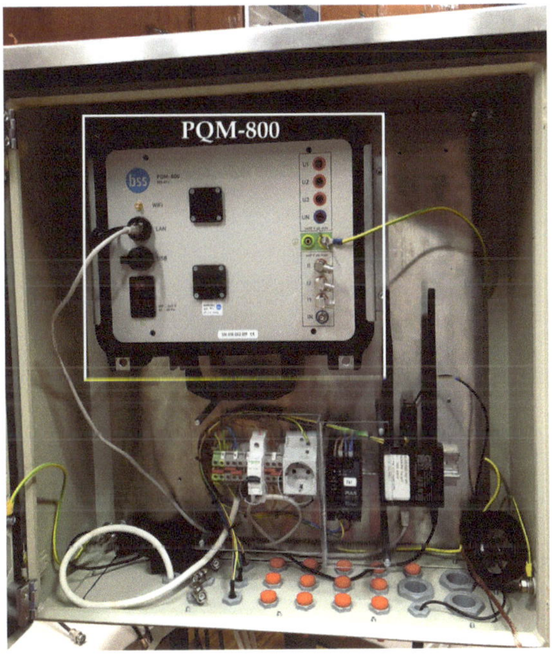

Figure 5. The control cabinet with the power quality monitor.

3. Tests in the High-Voltage Laboratory

Three CDSs are assembled for monitoring transient overvoltages in the substation. Prior to installation, the devices are tested in the high-voltage laboratory of the University of Stuttgart, where the following tests are conducted:

- Measurement of C_2 capacitances;

- Accuracy assessment (measurement of divider ratio and stray capacitances C_1);
- Lightning impulse test;
- Measurement of the CDS bandwidth.

The capacitances C_2 of the CDSs, formed by temperature-stable multilayer ceramic capacitors (MLCC), are measured using a multimeter. The values obtained for the devices MD1, MD2, and MD3 are 49.94 nF, 50.76 nF, and 50.53 nF, respectively.

3.1. Accuracy Assessment Using AC Voltage

The accuracy of the three CDSs is assessed by measuring the voltage ratio and the stray capacitance C_1 of the setup built in the high-voltage laboratory. Despite differences from the configuration in Figure 2, it is still possible to evaluate the linearity of the measurement system response since C_1 is the only parameter that changes based on the layout. The measurements are performed individually with an input voltage range from 13 kV_{rms} to 76 kV_{rms}. Figure 6 shows the schematic diagram (a) and the setup (b) for measuring the accuracy of the CDSs.

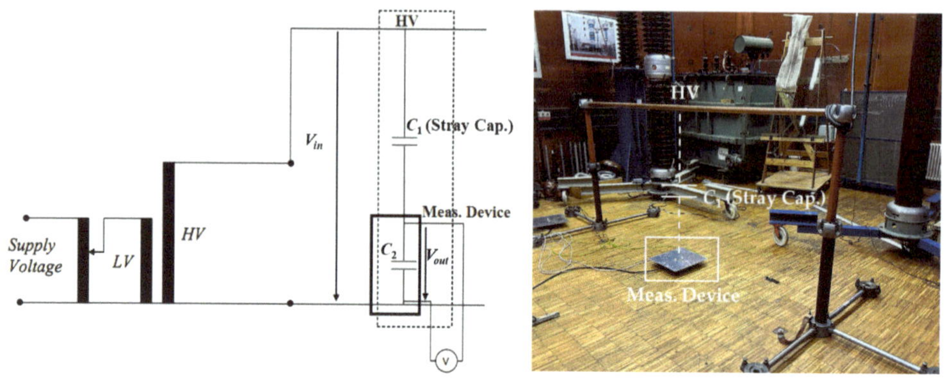

(**a**) (**b**)

Figure 6. Measurement of the accuracy of the CDSs. (**a**) Schematic diagram; (**b**) test setup.

The input high voltage V_{in} is measured using a standard capacitive divider with a ratio of 3060:1. Both signals, the output of the standard divider and the output of the CDSs (V_{out}), are evaluated with a 70 MHz oscilloscope with a sampling rate of 1 GS/s. The stray capacitance C_1 is then calculated from Equation (1) using the measured values of C_2, V_{in}, and V_{out}. Figure 7 shows the measured values of voltage ratio (a) and stray capacitance C_1 (b). Ten measurements are conducted for each device, with the input voltage ranging from 13 kV_{rms} to 76 kV_{rms}.

The median voltage ratios for devices MD1, MD2, and MD3 are 69,892:1, 71,503:1, and 70,574:1, respectively. The average voltage ratios are 69,859:1, 71,535:1, and 70,610:1. Distinct voltage ratios are expected due to the dependence of the ratio on C_1 and C_2 and the different capacitances C_2 among the CDSs.

The stray capacitance C_1 should ideally have a unique value for a given configuration. However, variations are observed in the measurements of the three units, potentially attributed to sensor sensitivity or slight misalignment of the devices during measurements. Considering the average value of C_1 for each unit, the maximum error observed is 0.29% for devices MD1 and MD2, and 0.46% for device MD3.

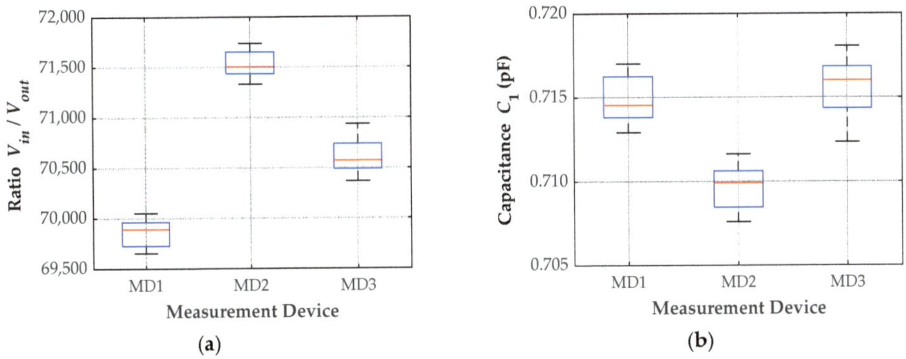

Figure 7. Measurement results. (**a**) Voltage ratio; (**b**) stray capacitance C_1.

The accuracy of the measurement devices is assessed by analyzing the voltage ratio measured for each device. The ratio error is calculated as follows:

$$\varepsilon = \frac{(k_r\, V_{out} - V_{in})}{V_{in}} \times 100\% \tag{2}$$

where k_r is the average of the measured voltage ratios for each device. The results are given in Table 1, which shows a linear behavior of the measurement devices concerning the applied voltage. The maximum error observed is 0.29% for devices MD1 and MD2, and 0.46% for device MD3, relative to their average voltage ratios.

Table 1. Ratio error considering the average of the measured voltage ratios.

Device MD1			Device MD2			Device MD3		
Input Voltage V_{in} (kV)	Meas. Voltage $k_{r\text{-}MD1}\, V_{out}$ (kV)	Ratio Error ε (%)	Input Voltage V_{in} (kV)	Meas. Voltage $k_{r\text{-}MD2}\, V_{out}$ (kV)	Ratio Error ε (%)	Input Voltage V_{in} (kV)	Meas. Voltage $k_{r-MD3}\, V_{out}$ (kV)	Ratio Error ε (%)
13.92	13.90	−0.15	13.16	13.16	0.03	13.46	13.49	0.17
20.20	20.19	−0.03	20.69	20.75	0.29	20.72	20.69	−0.13
28.09	28.01	−0.28	28.61	28.69	0.26	27.82	27.82	0.02
35.80	35.91	0.29	34.58	34.48	−0.28	35.19	35.30	0.33
41.31	41.43	0.28	41.31	41.20	0.26	41.31	41.38	0.16
47.12	47.08	−0.08	47.43	47.50	0.15	47.43	47.59	0.34
53.55	53.65	0.19	55.39	55.30	−0.16	54.47	54.37	−0.18
60.59	60.50	−0.15	62.12	62.02	−0.16	62.42	62.14	−0.46
67.63	67.62	0.00	68.54	68.60	0.08	68.85	68.63	−0.32
76.19	76.15	−0.06	76.50	76.54	0.05	76.19	76.26	0.08

3.2. Lightning Impulse Test

The lightning impulse test is the next evaluation performed in the high-voltage laboratory. The standard IEC 60060-1 [40] defines a lightning impulse as a double exponential waveform characterized by a front time T_1 of 1.2 µs ±30% and a tail time T_2 of 50 µs ±20%. The impulse generator can deliver switching and lightning impulse voltages up to 1 MV and a total energy of 30 kJ. Figure 8 illustrates the schematic diagram (a) and test setup (b).

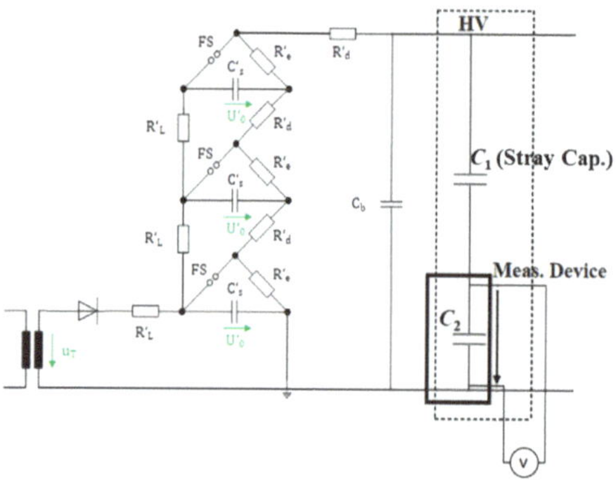

(a)

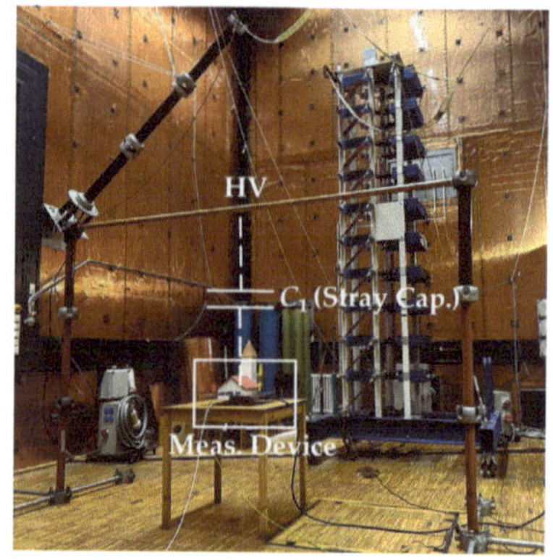

(b)

Figure 8. Lightning impulse test. (**a**) Schematic diagram; (**b**) test setup.

The applied impulse voltage is measured using a standard capacitive divider. Its output and the output from device MD1 are evaluated using an oscilloscope with a bandwidth of 1 GHz and a sampling rate of 10 GS/s. Figure 9 shows the measurement of the lightning impulse, where the output voltage from unit MD1 is multiplied by the voltage ratio measured for this configuration, which is 20,434:1.

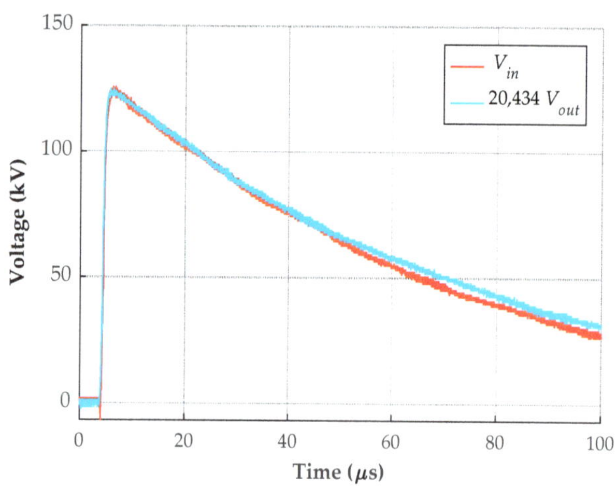

Figure 9. Lightning impulse response of device MD1.

The applied lightning impulse has a front time of 1.22 μs and a tail time of 51.48 μs, whereas device MD1 exhibits a front time of 1.17 μs and a tail time of 54.35 μs. Both signals meet the criteria of IEC 60060-1 and demonstrate similarity. Therefore, the measurement device can accurately reproduce a lightning impulse signal.

3.3. Measurement of the CDS Bandwidth

The measurement of the CDS bandwidth is performed using a 120 pF ceramic capacitor to simulate the stray capacitance C_1. Although the typical stray capacitance is less than 1 pF, a higher value is chosen to improve the signal-to-noise ratio of the measurement. Using a Rohde & Schwarz ZVRE vector network analyzer (VNA) [41] with a frequency range from 9 kHz to 4 GHz, the measurement assessed the V_{out}/V_{in} ratio over a frequency range spanning from 10 kHz to 10 MHz. Figure 10 shows the test setup (a) and the obtained measurement results (b).

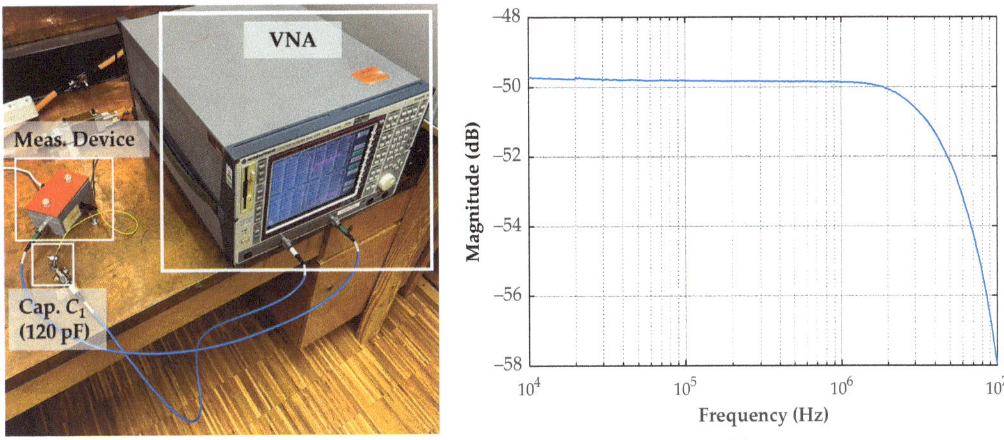

(**a**) (**b**)

Figure 10. Bandwidth measurement of device MD1. (**a**) Test setup; (**b**) measurement result.

The frequency response measurement shows a constant ratio of −49.8 dB up to 1.2 MHz, with the 3 dB frequency at 5.6 MHz. It indicates that the measurement device has a broadband response and can measure fast transients. However, the power quality monitor limits the bandwidth of the measurement system to a frequency range of up to 500 kHz, which is still sufficient to measure fast transients.

4. Installation and Calibration

The measurement system was installed in a high-voltage substation for a one-year measurement campaign. The CDSs were mounted on the bases of a three-phase disconnector in a transmission line bay. Devices MD1, MD2, and MD3 were installed in phases A, B, and C, respectively. The control cabinet was placed on the concrete base of a circuit breaker, approximately six meters from the devices. Figure 11 shows a CDS (a) and the control cabinet (b) installed in the substation.

The value of the stray capacitance C_1 depends on the distance and layout. Additionally, there is a coupling between the adjacent phases and the sensor, influencing its output voltage V_{out}, as illustrated in Figure 12.

Figure 12 shows only the coupling capacitances between the CDS installed in phase A and the three high-voltage connections. The same happens with the CDSs installed in phases B and C. To mitigate cross-coupling effects in the measurement, it is necessary to calibrate the measurement system. Two calibration techniques are examined:

- Simplified method (SM);
- Compensation of coupling capacitances (CCC).

4.1. Calibration Using the Simplified Method

The first calibration method compares the voltage values measured by the CDSs with the AC voltage from a substation divider at 50 Hz. In this case, the primary voltage measured by a capacitive voltage transformer serves as the reference. During the installation, the RMS value of the primary phase voltage was 238.45 kV, assuming a balanced three-phase system. The RMS voltages measured by the CDSs for phases A, B, and C were 2.03 V, 1.68 V, and 1.89 V, respectively. Therefore, the calculated voltage ratios for devices MD1, MD2, and MD3 are 117,461:1, 141,932:1, and 126,162:1, respectively. The voltage ratios can be set directly in the power quality monitor. Figure 13 shows the reference primary voltage (V_{in}) and the measured voltage (V_{meas}) considering the calculated voltage ratio.

(a)

(b)

Figure 11. Measurement system installed in the substation. (**a**) CDS on the base of a disconnector; (**b**) Control cabinet on the concrete base of a circuit breaker.

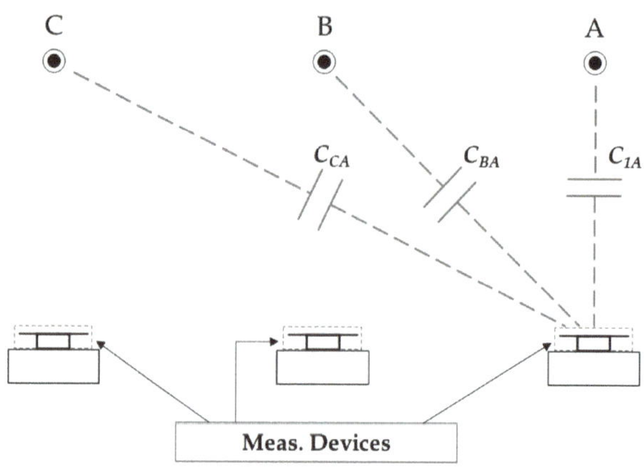

Figure 12. Coupling capacitances between a CDS and adjacent phases.

In phase B, the reference and measured voltages are in phase. It happens due to the symmetrical influence of phases A and C on the device installed in phase B. However, the measured voltage of phase A is delayed by 14.14° with respect to the reference voltage of the same phase due to the influence of phases B and C. Similarly, the measured voltage of phase C leads the reference voltage by 13.12° due to the influence of phases A and B.

This calibration method is generally suitable for measuring steady-state voltages. It is straightforward to implement directly in the power quality monitor. However, the coupling between the measurement devices and adjacent phases may introduce notable errors when measuring electromagnetic transients, as will be shown in Section 5.

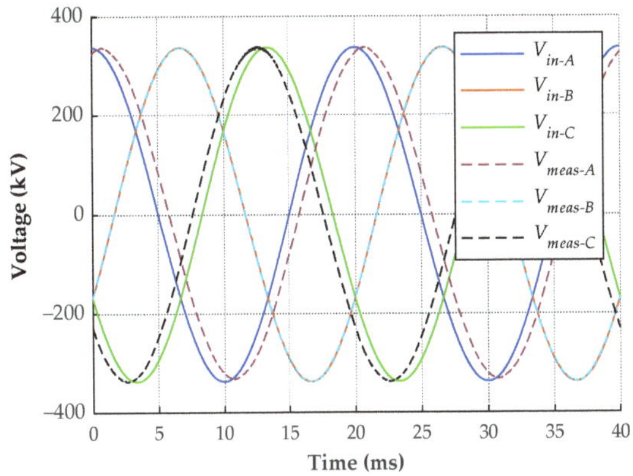

Figure 13. Comparison between the reference and measured voltages.

4.2. Compensation of the Coupling Capacitances

Given the constraints of the simplified method, especially in measuring transient overvoltages, a post-processing calibration method is developed. This method considers the coupling capacitances between the sensors and the high-voltage connections of the three phases to reconstruct the primary voltages from the measured output voltages.

This calibration method also requires the information of the steady-state primary voltage during the installation of the measurement system. The same primary voltage measured by a CVT and the corresponding output signals of the CDSs are used. The primary voltages are considered symmetrical, with the same amplitude and a phase difference of 120°. Therefore, the primary phase voltages are expressed in phasor notation as:

$$\begin{aligned} V_{1A} &= V_p \angle 0^\circ \\ V_{1B} &= V_p \angle 120^\circ \\ V_{1C} &= V_p \angle -120^\circ \end{aligned} \quad (3)$$

where V_p is the peak voltage measured by the CVT, in this case 337.22 kV.

Given the schematic diagram shown in Figure 12, an equivalent circuit for the measurement device installed in phase A is established to calculate the output voltage as a function of the primary voltage. The resulting equivalent circuit is presented in Figure 14.

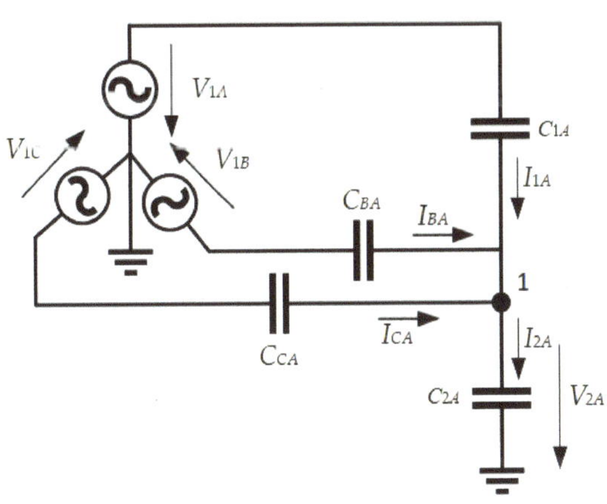

Figure 14. Equivalent circuit representing the measurement device installed in phase A.

Applying Kirchhoff's current law for node 1, the current I_{2A} can be calculated as:

$$I_{2A} = I_{1A} + I_{BA} + I_{CA} \tag{4}$$

where

$$\begin{aligned} I_{2A} &= j\omega C_{2A} \cdot V_{2A} \\ I_{1A} &= j\omega C_{1A} \cdot (V_{1A} - V_{2A}) \\ I_{BA} &= j\omega C_{BA} \cdot (V_{1B} - V_{2A}) \\ I_{CA} &= j\omega C_{CA} \cdot (V_{1C} - V_{2A}) \end{aligned} \tag{5}$$

Substituting the equations from (5) into Equation (4), the output voltage V_{2A} is calculated as:

$$V_{2A} = \frac{1}{(C_{2A} + C_{1A} + C_{BA} + C_{CA})}(C_{1A} \cdot V_{1A} + C_{BA} \cdot V_{1B} + C_{CA} \cdot V_{1C}) \tag{6}$$

The capacitance C_{2A} has a value of approximately 50 nF, which is much higher than the stray capacitances C_{1A}, C_{BA}, and C_{CA}, usually less than 1 pF. Therefore, Equation (6) can be simplified as:

$$V_{2A} = \frac{1}{C_{2A}}(C_{1A} \cdot V_{1A} + C_{BA} \cdot V_{1B} + C_{CA} \cdot V_{1C}) \tag{7}$$

The same concept and simplifications can be applied to the devices installed in phases B and C. Then, the output voltages are calculated as:

$$\begin{bmatrix} V_{2A} \\ V_{2B} \\ V_{2C} \end{bmatrix} = \begin{bmatrix} \frac{C_{1A}}{C_{2A}} & \frac{C_{BA}}{C_{2A}} & \frac{C_{CA}}{C_{2A}} \\ \frac{C_{AB}}{C_{2B}} & \frac{C_{1B}}{C_{2B}} & \frac{C_{CB}}{C_{2B}} \\ \frac{C_{AC}}{C_{2C}} & \frac{C_{BC}}{C_{2C}} & \frac{C_{1C}}{C_{2C}} \end{bmatrix} \begin{bmatrix} V_{1A} \\ V_{1B} \\ V_{1C} \end{bmatrix} \tag{8}$$

Equation (8) describes a linear system wherein the coefficient matrix depends on the stray capacitances and the user-defined capacitances C_2. Since the stray capacitances have a unique value for a particular layout, there is only one physical solution for this system. However, all stray capacitances are unknown, leading to nine unknown parameters. The symmetry in the sensor installation reduces the number of unknown parameters, yet even in this scenario, the system may not be entirely determined.

Assuming identical influences of phases A and C on the sensor installed in phase B, C_{AB} is equal to C_{CB}, and V_{2B} has the same phase as V_{1B} in steady-state voltage calibration. Additionally, the influence of phase B on the sensors in phases A and C is assumed to be the same, resulting in C_{BA} being equal to C_{BC}. The summarized assumptions are as follows:

$$C_{BA} = C_{AB} = C_{CB} = C_{BC} \tag{9}$$

As previously stated, the linear system has mathematically infinite solutions, even with the assumptions outlined in (9). One potential approach, explored in other works [30–32], is to neglect the coupling capacitance between the outer phases, i.e., to consider C_{CA} and C_{AC} equal to 0. However, this choice may result in substantial errors when evaluating transient overvoltages. An alternative is to assign values obtained from FEM simulation to these parameters. Upon defining C_{CA}, all parameters can be calculated by formulating the equations presented in (8), considering the assumptions given in (9).

The measurement system is calibrated using five distinct values of C_{CA}, ranging from 0 to 0.2 pF, to assess the impact of coupling between the outer phases. The values have been chosen so that the calculated parameters are of the same order of magnitude as those obtained in the FEM simulation. The calculated parameters are shown in Table 2.

Table 2. Set of stray capacitances calculated from different C_{CA} values.

Param. Set No.	C_{CA} (pF)	C_{AC} (pF)	C_{1A} (pF)	C_{1B} (pF)	C_{1C} (pF)	C_{BA} (pF)	C_{AB} (pF)	C_{CB} (pF)	C_{BC} (pF)
1	0.000	0.018	0.476	0.473	0.452	0.121	0.121	0.121	0.121
2	0.050	0.068	0.526	0.523	0.502	0.171	0.171	0.171	0.171
3	0.100	0.118	0.576	0.573	0.552	0.221	0.221	0.221	0.221
4	0.150	0.168	0.626	0.623	0.602	0.271	0.271	0.271	0.271
5	0.200	0.218	0.676	0.673	0.652	0.321	0.321	0.321	0.321

A transient overvoltage is reconstructed using the five parameter sets from Table 2. Figure 15 shows the signals from phase C, where the maximum overvoltage occurred.

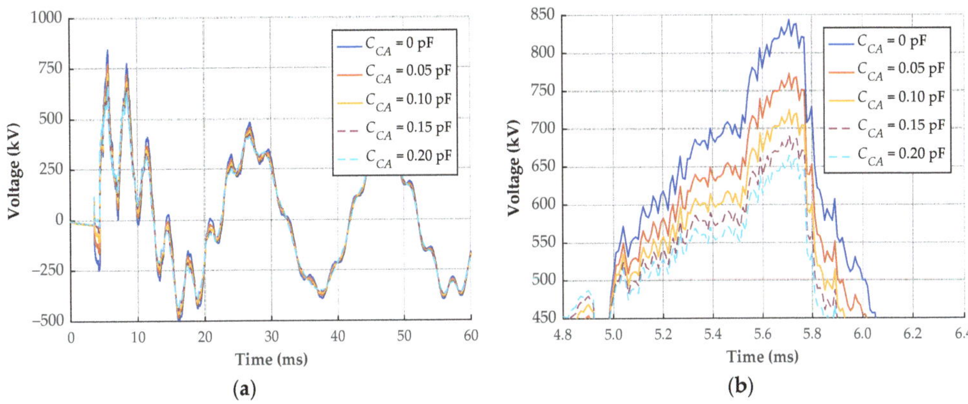

Figure 15. Transient overvoltage reconstruction with different calibration parameters. (**a**) Response of phase C; (**b**) zoomed waveforms on the region of maximum overvoltage.

In the scenario where C_{CA} is not considered (parameter set 1), the reconstructed signal exhibits a maximum overvoltage of 843 kV. Conversely, when C_{CA} is 0.20 pF (parameter set 5), the reconstructed signal reaches only 665 kV. This difference of 178 kV represents 21.1%. Therefore, neglecting the coupling capacitance between the outer phases may lead to substantial errors in reconstructing the primary voltage. The question that arises is how to compute the correct coupling capacitances.

Based on the established principle of parallel plate capacitors, some configurations yield a capacitance inversely proportional to the distance between the electrodes. Hence, the initial investigation aimed to establish this correlation for the specific layout where the measurement system was installed. However, this relationship could not be established.

Then, an investigation is conducted to determine if the stray capacitances exhibit an inverse proportionality to the square of the distance, represented as follows:

$$\begin{aligned} C_{1A} &\propto \frac{1}{{d_{AA}}^2} \\ C_{BA} &\propto \frac{1}{{d_{BA}}^2} \\ C_{CA} &\propto \frac{1}{{d_{CA}}^2} \end{aligned} \tag{10}$$

where d_{AA}, d_{BA}, and d_{CA} are the distances from the CDS installed in phase A to the high-voltage connections of phases A, B, and C, respectively.

Based on the information presented in (10) and considering that the permittivity and electrode area are identical for all phases, three normalized ratios can be calculated:

$$\begin{aligned} \frac{C_{1A}}{C_{BA}} &= \left(\frac{d_{BA}}{d_{AA}}\right)^2 \therefore \frac{C_{1A}}{C_{BA}} \cdot \left(\frac{d_{AA}}{d_{BA}}\right)^2 = 1 \\ \frac{C_{BA}}{C_{CA}} &= \left(\frac{d_{CA}}{d_{BA}}\right)^2 \therefore \frac{C_{BA}}{C_{CA}} \cdot \left(\frac{d_{BA}}{d_{CA}}\right)^2 = 1 \\ \frac{C_{1A}}{C_{CA}} &= \left(\frac{d_{CA}}{d_{AA}}\right)^2 \therefore \frac{C_{1A}}{C_{CA}} \cdot \left(\frac{d_{AA}}{d_{CA}}\right)^2 = 1 \end{aligned} \tag{11}$$

If the assumption that the capacitance is inversely proportional to the square of the distance is accurate, a set of stray capacitances satisfying the equations in (11) can be identified. Figure 16 shows the normalized ratios for different parameters, represented by the stray capacitance C_{CA}.

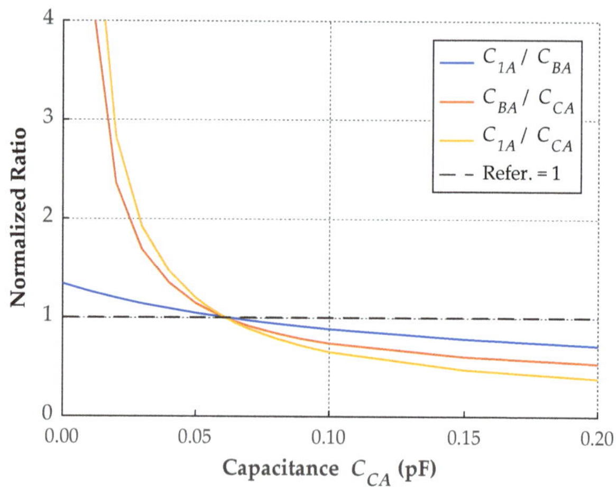

Figure 16. Normalized ratios according to different values of stray capacitances.

When the capacitance C_{CA} is 0.061 pF, the three normalized ratios become equal to 1. It indicates that a specific set of parameters satisfies the three equations given in (11). The parameters for the layout of this case study are presented in Table 3.

Table 3. Set of stray capacitances calculated for the specific layout.

C_{CA} (pF)	C_{AC} (pF)	C_{1A} (pF)	C_{1B} (pF)	C_{1C} (pF)	C_{BA} (pF)	C_{AB} (pF)	C_{CB} (pF)	C_{BC} (pF)
0.061	0.079	0.537	0.534	0.513	0.182	0.182	0.182	0.182

The correlation between the stray capacitances and the square of the distance has been demonstrated for the layout where the measurement system was installed. However, further investigation is necessary to confirm its validity for other configurations. If this correlation cannot be clearly established, alternative solutions include neglecting the coupling capacitance between the outer phases or estimating it using FEM simulations.

5. Measurement Results

The measurement system has been installed in the substation for one year, recording transients associated with transmission line switching and other disturbances. This section outlines the reconstruction of three transient signals using the calibration methods discussed in Section 4 and provides a statistical analysis of the recorded transients.

5.1. Comparison of Calibration Methods

As detailed in Section 4.1, the calibration provided by the simplified method is appropriate for measuring steady-state voltages. However, it may lead to significant errors when reconstructing transient signals. Therefore, the recommended approach is to utilize the method outlined in Section 4.2, which compensates for coupling capacitances. Nevertheless, calculating the coupling capacitances can be challenging, and a simplification can be made by considering the coupling capacitance between the outer phases as neglectable.

To assess the dissimilarities in the reconstruction of real transient signals, three transients are evaluated using the following methods and parameters:

- Simplified method (SM);
- Compensation of coupling capacitances (CCC) with C_{CA} = 0 pF;
- Compensation of coupling capacitances (CCC) with C_{CA} = 0.061 pF.

The first transient (also shown in Section 4) occurs due to the energization of the transmission line. Figure 17 shows the reconstructed signals for phase C (most critical).

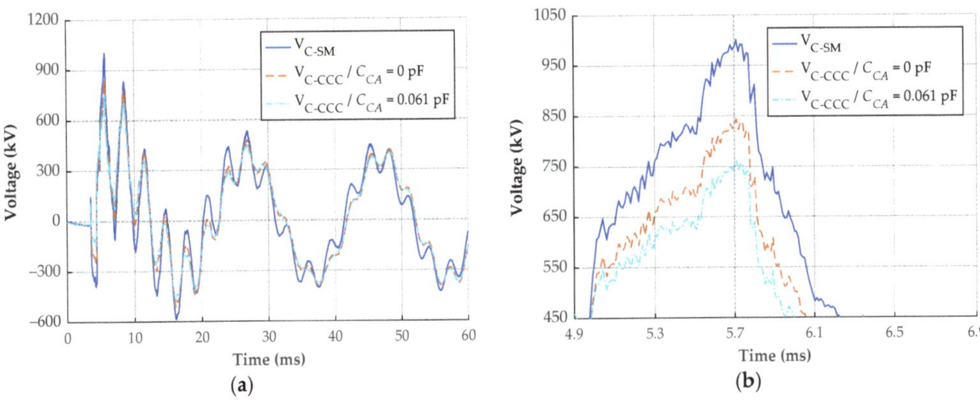

Figure 17. Transient overvoltage reconstruction—line energization. (**a**) Response of phase C; (**b**) zoomed waveforms on the region of maximum overvoltage.

The signal reconstructed using the CCC method with a C_{CA} of 0.061 pF is considered the reference. The maximum voltage of this signal is 761 kV. Ignoring the C_{CA} in the CCC method results in a maximum voltage of 843 kV, representing a difference of 82 kV or 10.8%. The signal reconstructed with the simplified method has a maximum voltage of 1000 kV, indicating a difference of 31.4%.

The second signal is also a transient that occurs during the energization of the transmission line. The signal reconstruction of phase C, where the maximum negative overvoltage happened, is shown in Figure 18.

Considering again the signal reconstructed using the CCC method with a C_{CA} of 0.061 pF as reference, the maximum negative voltage is −813 kV. Employing the CCC method with neglected C_{CA} results in a maximum negative voltage of −916 kV, representing a difference of 103 kV or 12.7%. The signal reconstructed using the simplified method has a maximum negative voltage of −1107 kV, representing a difference of 294 kV or 36.2%. Therefore, the simplified method evidently overestimates the maximum overvoltage of the analyzed transients, whereas the difference ranging from 10% to 13% when ignoring C_{CA} may or may not be considered acceptable, depending on the evaluation's objective.

The third evaluated transient was recorded during the de-energization of the transmission line. The reconstructed signals are shown in Figure 19.

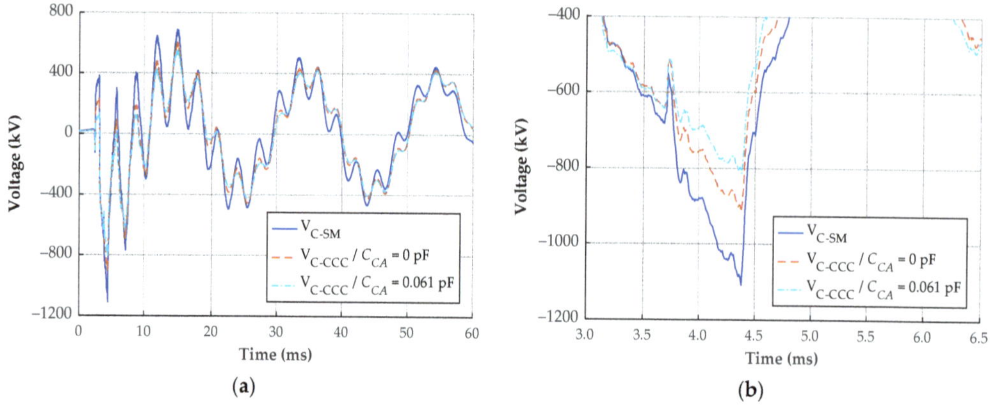

Figure 18. Transient overvoltage reconstruction—line energization. (**a**) Response of phase C; (**b**) zoomed waveforms on the region of maximum negative overvoltage.

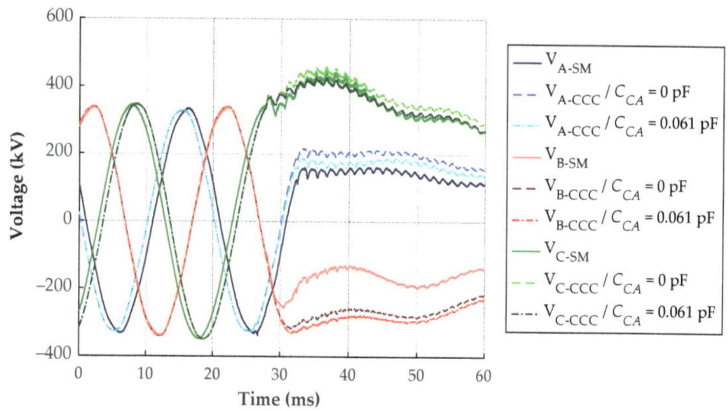

Figure 19. Transient overvoltage reconstruction—line de-energization.

The most significant difference between the reconstructed signals is observed in phase B. The simplified method indicates a difference of up to 148 kV compared to the CCC method with a C_{CA} of 0.061 pF. Moreover, the CCC method with neglected C_{CA} and CCC method with a C_{CA} of 0.061 pF have a maximum difference of approximately 30 kV. For this signal, no critical voltage level for equipment operation is identified. Consequently, the utilization of the simplified method would not compromise the evaluation.

5.2. Statistical Analysis of Transient Signals

During the one-year measurement campaign, the measurement system recorded 142 transient events. Table 4 shows the classification of these transients.

Table 4. Number of transient events recorded by the measurement system.

Line Energization (No. of Events)	Line De-Energization (No. of Events)	External Disturbance (No. of Events)	Total
27	28	87	142

Sixty-one percent of all recorded transients originate from external disturbances, such as circuit switching in the same substation or other perturbations in the power system. These signals are transients that return to the previous steady-state within a few milliseconds, meaning that there is no change in the state (closed/open) of the transmission line. The transmission line underwent 28 de-energizations and 27 energizations during the one-year period. This difference occurs because the line was energized when the measurement system was installed and de-energized when the system was removed.

The most critical transient overvoltages occurred during the energization of the transmission line. Therefore, a statistical analysis of the maximum voltage of the recorded transients is performed and the results are presented in Figure 20.

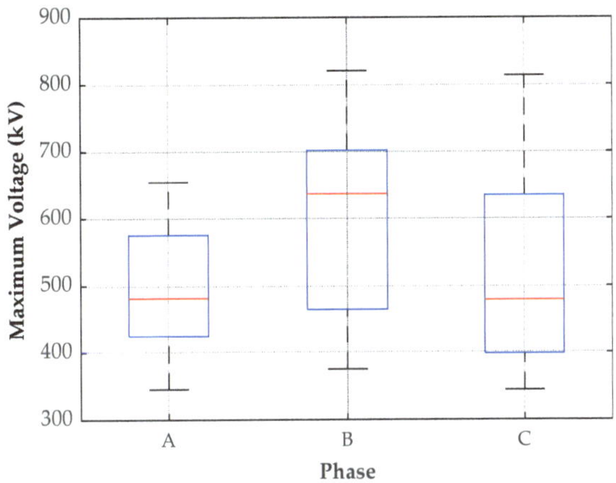

Figure 20. Maximum voltage measured during the energization of the transmission line.

The median of the maximum voltage registered for phases A, B, and C are 482 kV, 637 kV, and 479 kV, respectively. The maximum overvoltages are 654 kV, 820 kV, and 813 kV, respectively. Therefore, the maximum registered overvoltage, measured in phase B, represents about 2.4 times the rated peak voltage. Figure 21 shows the transient signal with the highest maximum (negative) overvoltage.

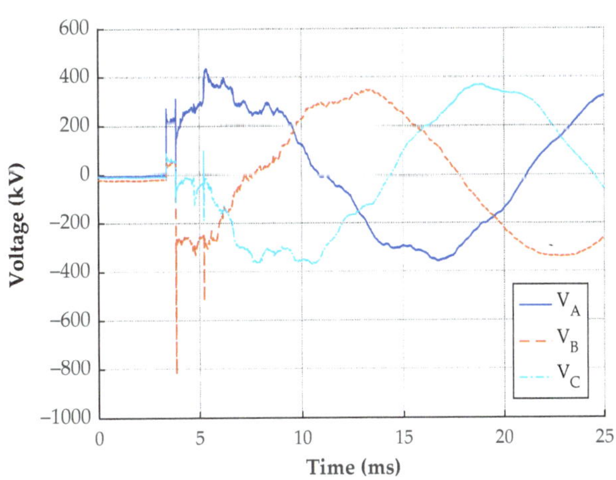

Figure 21. Transient signal with the maximum measured overvoltage.

6. Conclusions

Accurately measuring and investigating electromagnetic transients is gaining importance with the increasing integration of renewable energy sources into the power grid. This paper introduces a measurement system based on capacitive electric field sensors capable of accurately measuring fast transients due to its broadband response. The measurement system was installed in a high-voltage substation for one year, recording 142 transient events. The installation process is straightforward and does not require de-energizing the bay.

Calibration is essential for measuring transient overvoltages using the proposed measurement system. This paper discusses two techniques: the simplified method and the CCC method. The simplified method can measure steady-state voltages, but it has been proven inaccurate for measuring transients. On the other hand, the CCC method, which accounts for the coupling capacitances between the measurement devices and the three phases, is the most appropriate for reconstructing transient overvoltages. All coupling capacitances were successfully calculated based on the layout and simplifications. However, neglecting the coupling capacitance between the external phases is acceptable if no relation between the layout (distances) and the calculated capacitances can be established. This study found that neglecting it resulted in a difference between 10% and 13%.

Finally, the study demonstrates that the proposed measurement system serves as an efficient and flexible solution for the long-term monitoring of transient overvoltages in a high-voltage substation, especially for fast transients requiring broadband measurement.

Author Contributions: Conceptualization, M.B. and S.T.; methodology, F.L.P. and S.T.; software, F.L.P.; validation, F.L.P. and S.T.; investigation, F.L.P.; resources, M.B. and S.T.; writing—original draft preparation, F.L.P.; writing—review and editing, M.B. and S.T.; visualization, F.L.P.; supervision, M.B. and S.T. All authors have read and agreed to the published version of the manuscript.

Funding: This research received no external funding.

Institutional Review Board Statement: Not applicable.

Informed Consent Statement: Not applicable.

Data Availability Statement: Data are contained within the article.

Acknowledgments: The authors gratefully acknowledge the support of TransnetBW GmbH. Special appreciation is expressed to Kai A. Alsdorf for his significant contributions.

Conflicts of Interest: The authors declare no conflicts of interest.

References

1. *CIGRE TB 577A*; Electrical Transient Interaction between Transformers and the Power System—Part 1: Expertise. JWG A2/C4.39; CIGRE: Paris, France, 2014.
2. *CIGRE TB 577B*; Electrical Transient Interaction between Transformers and the Power System—Part 2: Case Studies. JWG A2/C4.39; CIGRE: Paris, France, 2014.
3. Probst, F.L.; da Luz, M.V.F.; Tenbohlen, S. Modeling of a capacitive voltage transformer for evaluation of transient response in conformity with the IEC 61869-5 Standard. *Electr. Power Syst. Res.* **2023**, *223*, 109591. [CrossRef]
4. *CIGRE TB 881*; Electromagnetic Transient Simulation Models for Large-Scale System Impact Studies in Power Systems Having a High Penetration of Inverter-Connected Generation. WG C4.56; CIGRE: Paris, France, 2022.
5. *IEC 60071-1*; Insulation Co-Ordination—Part 1: Definitions, Principles and Rules, Ed. 9.0. International Electrotechnical Commission: Geneva, Switzerland, 2019.
6. *IEC 60071-2*; Insulation Co-Ordination—Part 2: Application Guidelines, Ed. 4.0. International Electrotechnical Commission: Geneva, Switzerland, 2018.
7. Vermeulen, H.; Davel, P. Voltage harmonic distortion measurements using capacitive voltage transformers. In Proceedings of the IEEE AFRICON 1996, Stellenbosch, South Africa, 27 September 1996; Volume 2, pp. 1012–1017.
8. Meyer, J.; Stiegler, R.; Klatt, M.; Elst, M.; Sperling, E. Accuracy of harmonic voltage measurements in the frequency range up to 5 kHz using conventional instrument transformers. In Proceedings of the 21st International Conference on Electricity Distribution (CIRED 2011), Frankfurt, Germany, 6–9 June 2011.

9. Buchhagen, C.; Fischer, M.; Hofmann, L.; Däumling, H. Metrological determination of the frequency response of inductive voltage transformers up to 20 kHz. In Proceedings of the 2013 IEEE Power & Energy Society General Meeting, Vancouver, BC, Canada, 21–25 July 2013; pp. 1–5. [CrossRef]
10. Tong, L.; Liu, Y.; Chen, Y.; Su, S.; Liang, P. A CVT based lightning impulse wave measuring method using convolutional neural network. In Proceedings of the 2021 IEEE 4th International Electrical and Energy Conference (CIEEC), Wuhan, China, 20–30 May 2021; pp. 1–6.
11. Zare, M.; Mirzaei, A.; Askarian Abyaneh, H. Designing a compensating electronic circuit to enhance capacitive voltage transformer characteristics. In Proceedings of the 9th Power Systems Protection and Control Conference (PSPC2015), Tehran, Iran, 11–14 October 2015; pp. 12–18.
12. Kang, Y.C.; Zheng, T.Y.; Choi, S.W.; Kim, Y.H.; Kim, Y.G.; Jang, S.I.; Kang, S.H. Design and evaluation of a compensating algorithm for the secondary voltage of a coupling capacitor voltage transformer in the time domain. *IET Gener. Transm. Distrib.* **2009**, *3*, 793–800. [CrossRef]
13. Zadeh, H.; Zuyi, L. A compensation scheme for CVT transient effects using artificial neural network. *Electr. Power Syst. Res.* **2008**, *78*, 30–38. [CrossRef]
14. De La Hoz, M.; Chacón, J.; Sørensen, D.A.; Larrinaga, U.Z.; Barón, C.R. Harmonic measurements in a capacitive voltage transformer: Improvement considering the transformer's design parameters. In Proceedings of the 2021 International Conference on Smart Energy Systems and Technologies (SEST), Vaasa, Finland, 6–8 December 2021; pp. 1–6.
15. Ghassemi, F.; Gale, P.; Cumming, T.; Coutts, C. Harmonic voltage measurements using CVTs. In Proceedings of the 2004 IEEE Power Engineering Society General Meeting, Denver, CO, USA, 6–10 June 2004; Volume 1, p. 482.
16. Fu, C.Z.; Si, W.R.; Zhao, D.D.; Huang, H.; Jin, H.; Zhao, W.B. Research on the method for transient overvoltage measurement using CVT within a capacitor C3. In Proceedings of the 2016 International Conference on Condition Monitoring and Diagnosis (CMD), Xi'an, China, 25–28 September 2016; pp. 1028–1030.
17. Stirl, T. Online monitoring of capacitance-graded bushing on power transformers. In Proceedings of the ETG Technical Seminar No. 97, Cologne, Germany, 9–10 March 2004; pp. 115–120. (In German).
18. Stirl, T.; Skrzypek, R.; Tenbohlen, S.; Vilaithong, R. On-line condition monitoring and diagnosis for power transformers their bushings, tap changer and insulation system. In Proceedings of the 2006 International Conference on Condition Monitoring and Diagnosis (CMD), Changwon, Republic of Korea, 2–5 April 2006.
19. Ma, G.M.; Li, C.R.; Quan, J.T.; Jiang, J. Measurement of VFTO based on the transformer bushing sensor. *IEEE Trans. Power Deliv.* **2011**, *26*, 684–692. [CrossRef]
20. Tenbohlen, S.; Kattmann, C.; Brugger, T. Power quality monitoring in power grids focusing on accuracy of high frequency harmonics. In Proceedings of the 2018 Cigré Session, Paris, France, 26–31 August 2018.
21. Kattmann, C.; Siegel, M.; Tenbohlen, S.; Bruger, T. Power quality measurement in high voltage grids using transformer bushings. In Proceedings of the VDE Hochspannungstechnik, Berlin, Germany, 12–14 October 2018. (In German).
22. Botha, M. Wideband Modelling of Capacitive Voltage Sensors for Open-Air Transmission Line Applications. Ph.D. Thesis, Electrical Engineering, University of Stellenbosch, Stellenbosch, South Africa, 2002.
23. Liu, H.; Ma, G.; Qin, W.; Wang, W.; Du, H.; Wang, L.; Zhai, S.; Zeng, F.; Han, W. A low-cost and high-performance self-trigger method for high-voltage transient measurement. *High Volt.* **2022**, *8*, 412–420. [CrossRef]
24. Han, R.; Yang, Q.; Sima, W.; Zhang, Y.; Sun, S.; Liu, T.; Chen, S. Non-contact measurement of lightning and switching transient overvoltage based on capacitive coupling and pockels effects. *Electr. Power Syst. Res.* **2016**, *139*, 93–100. [CrossRef]
25. Sun, S.; Ma, F.; Yang, Q.; Ni, H.; Bai, T.; Ke, K.; Qiu, Z. Research on non-contact voltage measurement method based on near-end electric field inversion. *Energies* **2023**, *16*, 6468. [CrossRef]
26. Xiao, D.; Xie, Y.; Ma, Q.; Zheng, Q.; Zhang, Z. Non-contact voltage measurement of three-phase overhead transmission line based on electric field inverse calculation. *IET Gener. Transm. Distrib.* **2018**, *12*, 2952–2957. [CrossRef]
27. Chen, K.-L.; Guo, Y.; Ma, X. Contactless voltage sensor for overhead transmission lines. *IET Gener. Transm. Distrib.* **2018**, *12*, 957–966. [CrossRef]
28. Si, D.; Wang, J.; Wei, G.; Yan, X. Method and experimental study of voltage measurement based on electric field integral with Gauss–Legendre algorithm. *IEEE Trans. Instrum. Meas.* **2020**, *69*, 2771–2778. [CrossRef]
29. Wang, J.; Ban, S.; Yang, Y. A differential self-integration D-dot voltage sensor and experimental research. *IEEE Sens. J.* **2015**, *15*, 3846–3852. [CrossRef]
30. Wouters, P.A.A.F.; Barakou, F.; Gargari, S.M.; Smit, J.; Steennis, E.F. Accuracy of switching transients measurement with open-air capacitive sensors near overhead lines. In Proceedings of the 2018 IEEE International Conference on High Voltage Engineering and Application (ICHVE), Athens, Greece, 10–13 September 2018; pp. 1–4. [CrossRef]
31. Barakou, F.; Wouters, P.A.A.F.; Gargari, S.M.; Smit, J.; Steennis, E.F. Merits and challenges of a differentiating-integrating measurement methodology with air capacitors for high-frequency transients. In Proceedings of the 2018 Cigré Session, Paris, France, 26–31 August 2018.
32. Barakou, F.; Steennis, F.; Wouters, P. Accuracy and reliability of switching transients measurement with open-air capacitive sensors. *Energies* **2019**, *12*, 1405. [CrossRef]
33. Brügger, T.H. Transient and harmonic measurements with unconventional instrument transformers. In Proceedings of the Stuttgarter Hochspannungssymposium, Stuttgart, Germany, 21–22 June 2022. (In German)

34. Probst, F.L.; Beltle, M.; Gerber, M.; Tenbohlen, S.; Alsdorf, K.A. Measurement of switching transient overvoltages with a capacitive electric field sensor. *Electr. Power Syst. Res.* **2023**, *223*, 109619. [CrossRef]
35. Probst, F.L.; Beltle, M.; Gerber, M.; Tenbohlen, S.; Alsdorf, K.A. Using capacitive electric field sensors to measure transient overvoltages: A case study. In Proceedings of the International Symposium on High Voltage Engineering (ISH), Glasgow, Scotland, 28 August–1 September 2023.
36. CST Studio Suite [Computer Software]. 2021. Available online: http://www.cst.com (accessed on 23 November 2023).
37. Linear Technology. Fast Settling, JFET Input Operational Amplifier. LT1122 Datasheet, Revision B February 2014. Available online: https://www.analog.com/media/en/technical-documentation/data-sheets/1122fb.pdf (accessed on 19 February 2024).
38. Traco Power. DC/DC Converter TVN 5WI Series, 5 Watt. TVN 5-4825WI Datasheet, Revision September 2023. Available online: https://www.tracopower.com/de/deu/tvn5wi-datasheet (accessed on 19 February 2024).
39. BSS Hochspannungstechnik. PQM-800 Power Quality Monitor. PQM-800 Datasheet. Available online: https://www.bss-hochspannungstechnik.de/pdf/Datasheet%20PQM-800.pdf (accessed on 23 November 2023).
40. *IEC 60060-1*; High-Voltage Test Techniques—Part 1: General Definitions and Test Requirements. International Electrotechnical Commission: Geneva, Switzerland, 2010.
41. Rohde & Schwarz. Vector Network Analyzer Family ZVR. VNA ZVRE Datasheet. Available online: https://www.testequipmenthq.com/datasheets/Rohde-Schwarz-ZVRE-Datasheet.pdf (accessed on 19 February 2024).

sensors

MDPI

Article

Method of Equivalent Error as a Criterion of the Assessment of the Algorithms Used for Estimation of Synchrophasor Parameters Taken from the Power System

Malgorzata Binek and Pawel Rozga *

Institute of Electrical Power Engineering, Lodz University of Technology, Stefanowskiego 20, 90-537 Lodz, Poland; malgorzata.binek@p.lodz.pl
* Correspondence: pawel.rozga@p.lodz.pl

Abstract: The development of digital techniques in control engineering leads to the creation of innovative algorithms for measuring specific parameters. In the field of electric power engineering these parameters may be amplitude, phase and frequency of voltage or current occurring in the analyzed electric grid. Thus, the algorithms mentioned, applied in relation to the quoted parameters, may provide precise and reliable measurement results in the electric grid as well as ensure better grid monitoring and security. Signal analysis regarding its identification due to the type of interference is very difficult because the multitude of information obtained is very large. In order to indicate the best method for determining errors in measuring synchronous parameters of the measured current or voltage waveforms, the authors propose in this paper a new form of one error for all testing functions, which is called an equivalent error. This error is determined for each error's value defined in the applicable standards for each of selected 15 methods. The use of the equivalent error algorithm is very helpful in identifying a group of methods whose operation is satisfactory in terms of measurement accuracy for various types of disturbances (both in the steady state and in the dynamic state) that may occur in the power grid. The results are analyzed for phasor measurement unit (PMU) devices of class P (protection) and M (measurement).

Keywords: equivalent error; PMU; RBF; DFT; synchrophasors; phase and amplitude estimation

Citation: Binek, M.; Rozga, P. Method of Equivalent Error as a Criterion of the Assessment of the Algorithms Used for Estimation of Synchrophasor Parameters Taken from the Power System. *Sensors* **2024**, *24*, 4619. https://doi.org/10.3390/s24144619

Academic Editors: Hossam A. Gabbar and Andrea Cataldo

Received: 22 March 2024
Revised: 9 July 2024
Accepted: 15 July 2024
Published: 17 July 2024

1. Introduction

The power system consists of devices which are responsible for the generation, transmission, distribution, storage, and utilization of electricity. The main task the power system must ensure is supplying the energy to consumers in a continuous and uninterrupted manner, simultaneously minimizing the financial costs. Therefore, the power system must meet technical requirements that will ensure safe use and the required quality and reliability in the supply of electricity. Nevertheless, ensuring stability in energy supply forces the system supervision by measuring the technology providing network monitoring and real-time system status assessment [1]. The power system under operation is vulnerable to disturbances that may spread in a short time and cover a large area of the system. Therefore, their elimination usually involves identifying the place of their occurrence, the type of disturbance and selected parameters of the disturbance's characteristic values. Continuous and reliable supervision over the operating status of the power system, in order to ensure power security, is carried out by many systems, including, first of all, the Wide Area Measurement System (WAMS), which has phasor measurement units (PMUs) located at selected points of the power system. At the point of installation, PMUs measure instantaneous values of current or voltage based on the measurements of the phasor angle, magnitude, frequency, and rate of change of frequency (ROCOF) [2]. The PMU devices are installed in the power stations, the choice of which depends on the availability of the measurements being able to be conducted in a given station. In most applications, phasor data are obtained from locations with large distances from the PMU, which is why communication connections and data concentrators play an important role in collecting the data. PMU devices are synchronized by GPS (Global Positioning System) usually at one-second intervals [2–4]. They can determine the instantaneous phase angle of the phasor as the angle of the phasor relative to the coordinate system rotating with the rated frequency of the power system, synchronized with the Universal Time Clock (UTC) from the GPS system and mark each estimated value with a time stamp from the GPS system. The quantities calculated and marked in this way are called synchrophasor parameters, while the measurements are called synchronous measurements. The synchronous measurements in the WAMS, performed under the supervision of GPS devices, generate a synchronization error resulting from the accuracy of the GPS system, which is approximately 1 μs. In the synchronous measurements, this error corresponds to an angle shift of 0.018° or 0.0314 rad, which approximately generates a measurement error of about 0.03% [5–7].

As was mentioned above, the sinusoidal nature of voltages and currents in electricity transmission in a dynamic power system causes significant distortions in voltage and current waveforms (including, for example, voltage or current fluctuations, waveform distortions, and increases in voltage or current values) [8–11]. Signal oscillations can contribute to the sudden load changes, switching phenomena in transmission lines and, finally, even the blackout event. Therefore, accurate measurement of currents and voltages with an appropriately fast response time ensures the proper functioning of the modern power system (even small oscillations resulting from the structure of the power system may lead to a wide area blackout [9]). Thus, an important task in terms of protection and control of the power system operation is to ensure reliable and precise measurements of network parameters and then analyze them in a proper, fast time manner. Without a doubt, the main criterion for this goal is the obtainment of a high measurement accuracy.

When it comes to the PMU devices, determining the phasor parameters is possible using various algorithms, wherein an innovative algorithms are most often used. Obviously, the applicable standards impose appropriate requirements in the form of permissible measurement error values of synchrophasor parameters for disturbances, taking into account the class P (Protection) and class M (Measurement) of PMU devices operation. This is performed to assess the effectiveness and suitability of the measurement algorithm used and to check its sensitivity to an interference in input signals in the form of higher harmonics, aperiodic components, out-of-band interferences, range of frequency, or changes of frequency. The possibility of using various measurement algorithms for estimation of the amplitude, angle, and frequency of the synchrophasor means that a large number of results are obtained during this task. The solutions associated with protection, control, and monitoring of electric power systems introduce a variety of the parameters, on the basis of which the above-mentioned research is realized [3].

The literature reports dealing with this problem are very extensive, because the multitude and diversity of solutions influence the choice of a calculation method that adapts to the selected task in a specific phasor application. Phasor parameters for PMU devices can be determined based on the use of different methods. The algorithms used in determining phasor parameters include, first of all, methods based on Fourier Transform (FT) [12–20], which are very popular due to the speed and accuracy of data processing and the simplicity of application. In determining the phasor parameters, the Monte Carlo (MC) method [21] is also applied in the analysis of real data. In turn, the Wavelet Transform (WT) method and the Hilbert Transform (HT) method are used to detect the disturbances: noise and harmonics [22]. In the literature, the Mean Squared Error (MSE) method can also be found to be suitable to solve the problem of determining the synchrophasor parameters, most often in the form of Weighted Least Squares (WLS) [23,24] or Recursive Least Square (RLS) in matrix calculations [25], the Phase Lock Loop (PLL) method [3,26], the Quadrature Filter (FQ) [1,5–7] or the Kalman Filter (KF) [27]. Currently, methods based on Artificial Neural Networks (ANNs) are becoming increasingly popular, including network models described already in 1987 by R. Lippmann: the Hopfield network or Hamming network [28], Convolutional Neural Network (CNN) used to identify events in extracts of complementary features [29], a Machine Learning (ML)-based method used for fault location [25], Radial Basic Function (RBF), Linear Neural Network (LNN), and Feedforward Neural Network (FNN), all presented in [1–4,9–11,30–32], or the physics-aware pruned neural network (P2N2), based on the topology of the physical network and used to design connections between different hidden layers of a neural network model in order to improve the accuracy of voltage estimation in the distribution system [30]. Methods combining classical algorithms with neural networks are gaining more and more popularity in analyzing phasor parameters. An example may be the Discrete Wavelet Transform (DWT) combined with Deep Neural Networks (DNNs). This is because they provide greater accuracy of calculations and quick feedback regarding the verification of a disturbance or its location in the power network [33]. As mentioned above, the measurement requirements have been specified in the standards for two different performance classes: class P and class M. The first one is used when the application requires a quick response, while the second one is used when the response speed is not critical and the measurement precision can be increased at its expense. Since the number of results to be analyzed is very large, choosing the correct algorithm to analyze them is a concern in this area. However, due to the fact that there are several dozen or even several hundreds of different algorithms which are able to be chosen, it may be difficult to analyze so many obtained data results.

Hence, in order to indicate a method that ensures the highest accuracy and reliability of measurement for many testing functions, mainly from the point of view of development of security technology and control process, the authors propose a new, innovative algorithm called the method of equivalent error. The first approach to present it was the authors' earlier work [4], where information on how to determine synchrophasor errors and the permissible values were specified. Herein, the work is focused on methods that meet the requirements for measurement errors in relation to the values of permissible errors specified in the applicable standards. Thus, the paper compares equivalent errors determined for six error values defined in [5–7] depending on the type of testing function (for 106 or 182 test functions for the classes P and M, respectively) for each of the fifteen selected methods. In other words, the aim of the paper is to present the method, called by the authors the method of equivalent error, which is necessary when analyzing a large number of results obtained from the algorithms of estimation of the amplitude, phase, and frequency of the synchrophasor, e.g., by PMU devices in the WAMS. Specifically, as an original approach, which has not been proposed yet, the authors define one equivalent error for all static and dynamic test functions excluding those functions with a step magnitude or functions with a step phase (functions from 1 to 98 for class P and functions from 1 to 174 for class M) and one equivalent error only for functions with a step magnitude or functions with a step phase (functions from 99 to 106 for class P and functions from 175 to 182 for class M). Based on the above, two equivalent errors determine the total error equivalent to clearly indicate the method, which provides the smallest total error equivalent for all test functions (which correspond to a disturbance that may appear in the power system).

The organization of the paper is as follows. Section 2 presents the requirements imposed on the estimation methods of phasor parameters of the signal model used as well as a new equivalent error algorithm. Section 3 presents the numerical results for the P and M class in relation to the analyzed population of the functions, while Section 4 discusses the simulation results. Conclusions are given in Section 5.

2. Methodology of the Studies

2.1. The Permissible Errors of Phasor Parameters

Phasor in relation to a current or a voltage signal is a vector rotating in the coordinate system with the electrical power system's nominal frequency. In this regard, a synchrophasor is a phasor with a UTC timestamp in every sample [5–7].

Standards [5–7] define, in general, two types of signals of static and dynamic types applied for PMU devices of the P (protection) and M (measurement) class. As was mentioned above, class P is used in the area of protection that requires a quick response and simultaneously permits less precise measurements, particularly when the basic waveform of the voltage or current signal has a frequency different from 50 Hz. In turn, the M class devices are applied in the situation when the most important issue is measurement accuracy and when a longer delay than for the P class is tolerable. Different sequences of the testing functions are proposed for each of the above mentioned classes, which were described in detail in [4]. For all of these functions, the errors must be estimated by the PMU devices, as per IEEE Standards [5–7].

The testing functions implemented in the authors' simulations refer to all types of disturbances described in [5–7] and were described in [4]. Additionally, for static and dynamic signals, IEEE Standards [5–7] impose requirements in the form of permissible errors. For the purposes of implementing their own method, called the equivalent error, the authors conducted appropriate tests that meet the requirements [5–7] of permissible errors for

(a) All testing functions excluding step functions:
  - Total vector error (TVE),
  - Frequency error (FE),
  - The rate of change of frequency error (RFE)

(b) All testing functions for step functions:
  - Response time of TVE (RT-TVE),
  - Delay time (DT),
  - Overshoot/Undershoot value (OV).

The definition of the above quantities (TVE, FE, RFE, RT-TVE, DT, OV) was presented in the authors' earlier article [4].

Definitions of the synchrophasor errors according to Standards [5–7], as well as the method of analyses of synchrophasor parameters for each functions, are described in detail in [4].

Appropriate tests for the selected 15 synchrophasor parameter estimation methods, taking into account the limitations for the obtained errors specified in the standards [5–7], were carried out for one continuous signal successively containing all testing functions, where one testing function lasts 1 s of the waveform analyzed. The exceptions are dynamic functions with a positive and negative ramp frequency, where the frequency range is ±2 Hz for P class and the frequency changes (increases/decreases) for 4 s.

In addition, at the beginning of testing, a signal with a frequency value equal to the value of the initial frequency will be fed for 0.4 s, and a signal with a frequency value equal to the value of the final frequency will be given for 0.6 s at the end of testing. The total duration of the signal is 5 s.

However, in the case of the M class, where the frequency range is ±5 Hz, the frequency will change (increase/decrease) for 10 s. Additionally, at the beginning of the test, a signal with a frequency value equal to the initial frequency will be given for 0.4 s, and at the end of the test, for 0.6 s a signal with a frequency value equal to the value of the final frequency will be given. In total, this signal will last 11 s.

2.2. *The Estimation Algorithm of the Phasor Parameters*

A new algorithm for determining the total equivalent error for each of the fifteen analyzed methods, in accordance with the procedure described in [4], was applied in the studies. The equivalent error algorithm proposed by the authors aims to determine, for each of the six synchrophasor errors (TVE, FE, RFE, RT_TVE, DT, OV) obtained according to the procedure explained in [4], the equivalent synchrophasor error (TVE_z, FE_z, RFE_z, RT_TVE_z, DT_z, OV_z) for each of the fifteen tested methods. Based on the obtained results, the total equivalent error is determined, indicating the method that results in the smallest total equivalent error (Er_tot), defined in Section 3.

The algorithm can be used in class P and class M devices, because it meets the requirements imposed by applicable standards [5–7].

Testing of the algorithm is conducted by means of computer simulations using the Matlab program (version R2024a), and its aim is to calculate measurement errors which are compared with permissible errors and on this basis the equivalent errors are determined.

2.3. *The Phasor Magnitude and Phase Angle Estimation Algorithms*

With respect to the guidelines set out in the applicable standards [5–7], simulations for static and dynamic signals were performed. For the synchrophasor frequency estimation, the zero-crossing method was used (described in [3,4]) as one of the most frequently considered algorithms in terms of its calculation simplicity and accuracy.

Based on the literature analysis, the authors presented various measurement algorithms used for amplitude and phase estimation of the synchrophasor: Orthogonal Components [19] (denoted as method 1), Fast Fourier Transform (DFT) [12,33] (method 2), Orthogonal Components with a single delay equal 1 [19] (method 3), Orthogonal Components with a single delay equal 64 [19] (method 4), Orthogonal Components with a double delay equal 1 [19] (method 5), Orthogonal Components with a double delay equal 21 [19] (method 6), Correlation with sine/cosine functions [19] (method 7), Convolution with orthogonal functions [19] (method 8), Least Square Method [1–4,34] (method 9), Quadrature Filter [5–7] (method 10), DFT method for non-nominal frequency [12] (method 11), Phase Locked Loop [35] (method 12), and methods based on artificial neural networks for estimating the initial amplitude of the synchrophasor: RBF [1–4,11,32,36,37] (method 13), FNN [32,38] (method 14), and LNN [32,38] (method 15). These methods met all requirements regarding the accuracy of signal magnitude identification in the assumed range of frequency, phase, and size changes.

For each of the methods analyzed in the article, allowing with sufficient accuracy the magnitude, phase, and frequency of the synchrophasor, an algorithm was used to introduce corrections to the results of the initial estimation of the magnitude and phase (described in [4]).

For the purposes of testing in accordance with [5–7], the linear phase FIR filter was used, which introduces a constant group delay, so that the difference between the input and output signals is linearly related to the frequency of the input signal. In order to compensate the filter delay, appropriate correction factors [4] were implemented for the estimated output signal phase and compensation for the magnitude estimation obtained as a result of using ANN methods. In the case of amplitude, piecewise linear approximation with a step of 1 Hz was implemented.

The first ANN method—the RBF method, presented in [3,4,11]—is universally applicable in many applications due to its quick learning process, correct generalization of data, and relatively simple network structure [36,37].

The second ANN method—the Feedforward Neural Network (method 14)—is an example of a network which consists of multiple layers, where the first of these has a connection from the network input, each subsequent layer is a connected from the previous layer, and the last layer generates the network's output. The Feedforward Neural Network can be used for input to output mapping, but a more specialized feedforward network can found applicable in fitting and pattern recognition.

The third ANN method—the Linear Neural Network (method 15)—is a type of network which characterizes the linear transfer function; therefore, their output can be of any value. Most commonly is applicable to solve problems described linearly, because a neuron can be trained by linear approximation of a non-linear function [38]. The advantage of this network is the fact that the information obtained at the output is close to the expected goal.

The neural networks analyzed herein underwent the training process for the same signal values: constant angle values (equal to 0), frequency (50 Hz), and a magnitude varying from 0.1 pu to 2.0 pu, with a step of 0.1 pu.

In the tested RBF, FFN, and LNN networks, samples from the neighborhood of the extreme learning signal were used as the learning function. Then, after verifying whether the above conditions were met, network training was carried out. The number of the tests performed allows us to conclude that each of the three tested ANN types provide adequate accuracy. Additionally, it can be noticed that even when the ANN network has the angle and frequency values unchanged during training, the estimation for the input signal magnitudes with any frequency, amplitude, or phase parameters is correct. An important aspect is to retain as many samples as were in the learning process, which are from the learning signal neighborhood and are extreme for each tested input signal.

An additional condition that each of the three tested ANN methods meets is the use of 128 samples, which is equivalent to no more than half of each measurement window. However, the number of samples from the previous window is 30 for the analyzed functions with a frequency smaller than 50 Hz.

For the purposes of conducting appropriate tests, it was assumed that the reporting frequency will always be $F_S = 50$ frames/s. This means that the estimated parameters of the synchronizer should be obtained from the signal with a maximum length of 0.02 s regardless of the current network frequency. One waveform period is analyzed for a network frequency of 50 Hz or above and for a network frequency of less than 50 Hz the result should be obtained by analyzing less than one mileage period. Tests of algorithms for processing instantaneous values of current or voltage for the synchrophaser parameters and subsequent verification were carried out only with the help of computer simulations in the Matlab application [32].

When choosing the above quoted methods, the possibility of their implementation in a real power system was guided. The selected methods represent various families of algorithms in the time domain but have not been tested due to the property of amplifying noise or interference other than those described in the standard [5–7]. In the article, the phasor errors, TVE, FE, RFE, OV, DT, RT, for all 15 methods were determined in accordance with the guidelines in the IEEE C37.118.1 standard. The actual phasor values may differ from those obtained from PMU measurements in both amplitude and phase.

3. Analysis of the Results for TVE Value

The analyses considered 15 methods of estimating the synchrophaser parameters, and for each method, six error values defined in the respective standard are determined for 106 (class P) or 182 (class M) test functions. Analyzing such a large number of results in order to choose the best method is very difficult. As an example, the TVE value results for testing functions are presented for class P and class M (Figure 1).

For each class, four methods with the smallest TVE values were selected, i.e., three classic methods and one method based on neural networks. Because the results obtained for method 13 and method 15 are practically the same, method 13 (as the first) was selected for analysis.

From the comparison of the obtained results presented in Figure 1, it can be concluded that the smallest TVE errors for all testing functions described in [5–7] are provided by method 13 (RBF). For only functions with a step phase change of $\pm\pi/10$ and with a step occurring in the middle of the measurement window (denoted as 105 to 106 for the P class and functions 181 to 182 for the M class in Figure 1), the errors obtained by this method are large (amounting to respectively: 17.9192 and 16.9530 for both classes). From classic methods presented in this article and denoted as 1 to 12, the lowest TVE errors were obtained using method 2.

It is difficult to analyze the results for each of the 15 methods containing all tested functions and six types of estimation errors, TVE, FE, RFE, RT-TVE, DT, and OV, for each class (P and M); therefore, a new form of one error is defined, called an equivalent error.

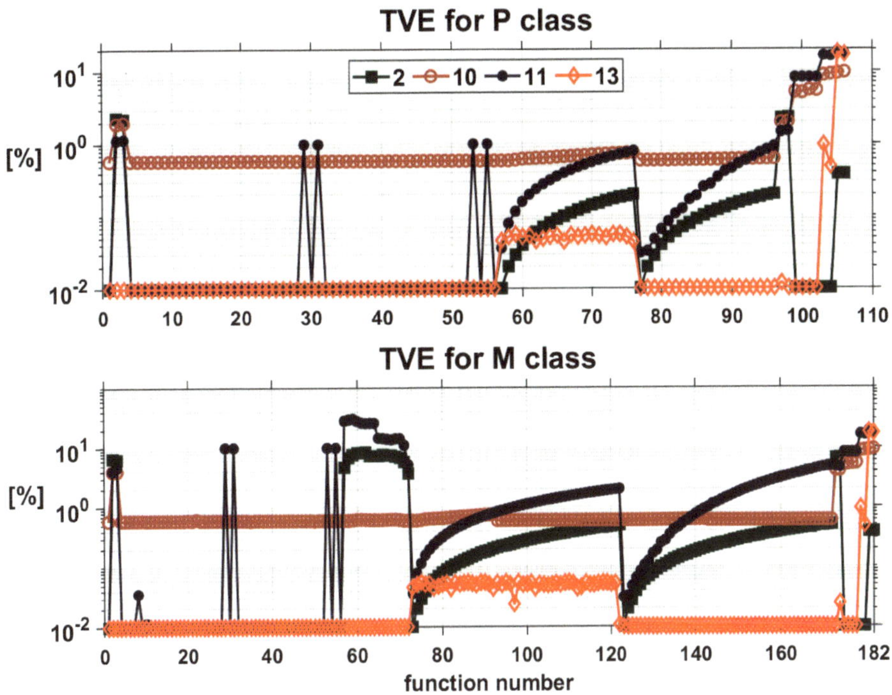

Figure 1. The TVE graphs for subsequent functions.

The definition of equivalent error for the TVE, FE, and RFE values for all functions except functions with a step magnitude or functions with a step phase, described in [5–7], are expressed in the following formulas:

$$TVE_z = \overset{174}{\sum_{j=1}^{98}} \left(\frac{TVE_j - TVE_{dj}}{TVE_{dj}} RT - TVE_j \right), \tag{1}$$

where TVE_j is the error TVE for the jth function, TVE_{dj} is the permissible error values TVE for the jth function, and $RT - TVE_j$ is the duration of magnitude, e.g., TVE, for the jth function above the permissible value.

$$FE_z = \overset{174}{\sum_{j=1}^{98}} \left(\frac{FE_j - FE_{dj}}{FE_{dj}} RT - FE_j \right), \tag{2}$$

where FE_j is the error FE for the jth function, FE_{dj} is the permissible error values FE for the jth function, and $RT - FE_j$ is the duration of magnitude, e.g., FE, for the jth function above the permissible value.

$$RFE_z = \overset{174}{\sum_{j=1}^{98}} \left(\frac{RFE_j - RFE_{dj}}{RFE_{dj}} RT - RFE_j \right), \tag{3}$$

where RFE_j is the error RFE for the jth function, RFE_{dj} is the permissible error values RFE for the jth function, and $RT - RFE_j$ is the duration of magnitude, e.g., RFE, for the jth function above the permissible value.

Based on the determination of the above equivalent errors, TVE_z (Equation (1)), FE_z (Equation (2)), and RFE_z (Equation (3)), a total equivalent error Er_1 (Equation (4)) has been determined for each method to clearly indicate which method generates the smallest errors

$$Er_1 = \sum\nolimits_j TVE_z + FE_z + RFE_z, \tag{4}$$

A similar procedure for determining equivalent errors for RT-TVE, DT, and OV errors defined by the standard [5–7] for functions with a step magnitude or functions with a step phase was used for all the considered methods:

$$RT - TVE_z = \sum\nolimits_{\substack{j=175\\ j=99}}^{\substack{182\\ 106}} \left(\frac{TVE_j - TVE_{dj}}{TVE_{dj}} RT - TVE_j \right), \tag{5}$$

$$DT_z = \sum\nolimits_{\substack{j=175\\ j=99}}^{\substack{182\\ 106}} DT_j, \tag{6}$$

where DT_j is the value DT for the jth function.

$$OV_z = \sum\nolimits_{\substack{j=175\\ j=99}}^{\substack{182\\ 106}} \left(\frac{OV_j - A_j}{A_j} RT - OV_j \right), \tag{7}$$

where OV_j is the value OV for the jth function, A_j is the maximum unit amplitude, which is 1.1 for overshoot or 0.9 for undershoot, the amplitude of the angle is $\pm\pi/18$, after step for the jth function, and $RT - OV_j$ is the time OV jth function more than the value A_j.

For $RT - TVE_z$ (Equation (5)), DT_z (Equation (6)), and OV_z (Equation (7)) equivalent errors, the total equivalent error Er_2 (Equation (8)) has the form

$$Er_2 = \sum\nolimits_j RT - TVE_z + DT_z + OV_z, \tag{8}$$

Based on Er_1 (Equation (4)) and Er_2 (Equation (8)) errors, a total equivalent error Er_{tot} (Equation (9)) was determined for all functions specified in [5–7] and each method individually:

$$Er_{tot} = \sum_j Er_1 + Er_2 \tag{9}$$

4. Equivalent Error Results

Three equivalent errors: TVE_z (Equation (1)), FE_z (Equation (2)), and RFE_z (Equation (3)), are identified for each of the 15 methods tested for all testing functions, excluding functions with a step magnitude or functions with a step phase. The results are presented in Table 1 for both class P and M.

Table 1. Equivalent error values for TVE, FE, RFE values.

Method Number	Equivalent Error					
	P Class			M Class		
	TVE_z	FE_z	RFE_z	TVE_z	FE_z	RFE_z
1	8.914	44.68	0.3348	427.9	1617	3.046
2	7.562	44.68	0.3348	168.7	1617	3.046
3	6.451	44.68	0.3348	421.4	1617	3.046
4	1.329	44.68	0.3348	383.0	1617	3.046
5	6.480	44.68	0.3348	422.0	1617	3.046
6	6.090	44.68	0.3348	427.0	1617	3.046
7	0.9379	44.68	0.3348	381.0	1617	3.046
8	0.9460	44.68	0.3348	381.2	1617	3.046
9	40.30	44.68	0.3348	629.5	1617	3.046
10	8.164	15.128	15.31	43.15	3.215	62.34
11	0.4843	39.19	0.6746	306.7	2849	8.223
12	574.3	46.13	10.76	2743	331.8	74.69
13	0.0000	0.0000	0.0000	0.0000	0.0000	0.0000
14	0.0000	0.0000	0.0000	0.0000	0.0000	0.0000
15	0.0000	0.0000	0.0000	0.0000	0.0000	0.0000

Based on the above results, it can be concluded that for both class P and class M, the smallest TVE_z, FE_z, and RFE_z errors were obtained for methods 13, 14, and 15, and the largest were obtained for method 12. Additionally, the TVE_z, FE_z, and RFE_z results for class M are several times higher compared to class P for classical methods (from 1 to 12). The results for methods 13, 14, and 15 remain unchanged.

For functions with a step magnitude or functions with a step phase, the results obtained by 15 methods tested for $RT - TVE_z$ (Equation (5)), DT_z (Equation (6)), and OV_z (Equation (7)) are shown in Table 2 for both classes. An analysis of the results obtained indicates that the lowest values of replacement errors are obtained using method 2 and methods 13, 14, and 15 for both classes. The results of replacement errors obtained for classes P and M are very similar.

Table 2. Equivalent error values for RT-TVE, DT, and OV values.

Method Number	Equivalent Error					
	P Class			M Class		
	$RT-TVE_z$	DT_z	OV_z	$RT-TVE_z$	DT_z	OV_z
1	9.041	0.3347	1.090	9.041	0.3347	1.090
2	0.0000	0.08000	0.0000	0.0000	0.08000	0.0000
3	8.473	0.3346	1.044	8.473	0.3346	1.044
4	8.626	0.3346	1.061284	8.626	0.3346	1.061
5	8.475	0.33467	1.044366	8.475	0.3346	1.044
6	8.551	0.3347	1.053475	8.551	0.3347	1.053
7	8.860	0.3364	1.090922	8.860	0.3364	1.091
8	8.860	0.3364	1.172316	8.860	0.3364	1.172
9	8.860	0.3377	1.069	8.858	0.3377	1.069
10	1.310	0.1681	0.0000	2.819	0.7762	0.0377
11	4.076	0.0008	0.4361	4.076	0.0008	0.4361
12	118.0	0.0008	3.877	118.0	0.0008	3.877
13	0.6574	0.1995	0.02048	0.6574	0.1995	0.02048
14	0.6592	0.1995	0.02048	0.6592	0.1995	0.02048
15	0.6574	0.1995	0.0205	0.6574	0.1995	0.0205

Because any type of disturbance may appear in the power grid, the total equivalent error Er_tot (Table 3) was defined for all 106 functions of class P and 182 functions of class M for each method. Additionally, the results of the Er_1 (for functions 1–98 for P class and 1–174 for M class) and Er_2 (for functions 99–106 for P class and 175–182 for M class) are presented in Table 3 presented for each method. Analyzing the obtained results, only neural network methods 13, 14, and 15, to a small extent, are susceptible to disturbances in the form of a step phase (described in [5–7]) that may appear in the power system.

Table 3. Value equivalent errors for Er_1, Er_2, and Er_tot for the P and M class of the PMU devices.

Method Number	Equivalent Error					
	P Class			M Class		
	Er_1	Er_2	Er_tot	Er_1	Er_2	Er_tot
1	53.93	10.47	64.39	2048	10.47	2059
2	52.57	0.0800	52.65	1789	0.0800	1749
3	51.46	9.852	61.32	2042	9.852	2051
4	46.34	10.02	56.36	2003	10.02	2013
5	51.49	9.854	61.35	2042	9.854	2052
6	51.10	9.939	61.04	2047	9.939	2057
7	45.95	10.29	56.24	2001	10.29	2012
8	45.96	10.37	56.33	2001	10.37	2012
9	85.31	10.27	95.57	2250	10.27	2260
10	38.59	1.479	40.07	108.7	3.632	112.3
11	40.35	4.513	44.86	3164	4.513	3168
12	631.2	121.9	753.1	3149	121.9	3271
13	0.0000	0.8774	0.8774	0.0000	0.8774	0.8774
14	0.0000	0.8792	0.8792	0.0000	0.8792	0.8792
15	0.0000	0.8774	0.8774	0.0000	0.8774	0.8774

5. Discussion and Conclusions

The paper presents a comparison of twelve methods, known from the literature, used in the estimation of the synchrophasor magnitude, phase, and frequency by PMU devices of the WAMS, with three methods based on artificial neural networks. In total, 15 different methods of synchrophasor parameter estimation were considered. All the calculations were carried out using the Matlab application. In order to indicate the usefulness of the method developed by the authors, simulations were performed for all testing functions in relation to disturbances defined in the applicable standards. This resulted in 106 testing functions for class P and 182 for class M. As a result, 25,920 error values were obtained for the selected functions, which did not allow for an unambiguous indication of the best method. Hence, it was decided to define own substitute errors for each error type, which reduced the number of errors to six. Finally, the optimal solution turned out to be the author's equivalent error method.

The research showed that the proposed method is characterized by the effectiveness, simplicity of the calculation process, accuracy, and acceptable speed of obtaining the answer. In order to indicate the potential of the equivalent error method, appropriate simulations were performed, and the results for the TVE_z, FE_z, and RFE_z values are presented in Table 1, while the results for the $RT\text{-}TVE_z$, DT_z, and OV_z values are presented in Table 2. In order to indicate the method generating the smallest error for the obtained results, the error was divided into the equivalent error Er_1 (the sum of errors TVE_z, DT_z, and OV_z) and the equivalent error Er_2, which is the sum of errors $RT\text{-}TVE_z$, DT_z, and OV_z. Based on these two errors, the total equivalent error Er_tot was determined for each method, and the results are presented in Table 3. The effectiveness of the solutions obtained using the equivalent error method for functions excluding those functions with a step magnitude or functions with a step phase is optimal in terms of methods 13, 14, and 15, as the methods generating the smallest errors, and method 12, generating the highest error values (Tables 1 and 3). For functions with a step magnitude or functions with a step phase, the interpretation of the results in Tables 2 and 3 indicates method 2 as generating the lowest error values. For this group of functions, methods 13, 14, and 15 provide higher error values compared to method 2, which only proves their sensitivity to this type of disturbance, but the values of equivalent errors obtained by them do not exceed 1%.

Defining the total equivalent error (Er_tot) was intended to uniquely indicate, among the 15 methods tested, the method (or group of methods) that provides the lowest error value. The analysis of the results presented in Table 3 clearly indicates that only methods 13, 14, and 15 generate a total equivalent error not exceeding 1%. Also, the differences in Er_1, Er_2, and Er_tot obtained using the methods based on artificial neural networks are negligibly small.

Thus, the form of the total equivalent error proposed in the paper may be stated to be successfully used for the following applications:

- In PMU devices applied in the power system (due to the reliability and precision of the results of the analysis of the measurements of network parameters, mainly current and voltage) for the purposes of developing protection technologies and control processes; in the optimization of the safety margin of system operation; or in the estimation of system operation or protection algorithms in the transmission networks,
- As a criterion for selecting a method that would work equally well for measurements made at the beginning and at the end of a long line, because it eliminates higher harmonics well from the input signals and estimates the amplitude and phase parameters with sufficient accuracy in the case of functions with a magnitude modulation and the frequency changing in a specific range, or in the case of the functions with a phase modulation and the frequency changing in a specific range, gives good responses for any signal starting angle.

On the basis of the obtained values of the total equivalent error, it can be concluded that the smallest total equivalent error for the M and P class functions was obtained for algorithms based on ANN—the value of the total equivalent error was the same for all the three methods analyzed. Thus, each of them is similarly useful in estimating the synchrophasor parameters used in the PMU devices of the WAMS.

Author Contributions: Conceptualization, M.B. and P.R.; methodology, M.B.; software, M.B.; validation, M.B. and P.R.; formal analysis, M.B. and P.R.; investigation, M.B.; data curation, M.B.; writing—original draft preparation, M.B. and P.R.; writing—review and editing, M.B. and P.R.; visualization, M.B. All authors have read and agreed to the published version of the manuscript.

Funding: This research received no external funding.

Institutional Review Board Statement: Not applicable.

Informed Consent Statement: Not applicable.

Data Availability Statement: The data are contained within the article.

Conflicts of Interest: The authors declare no conflicts of interest.

References

1. Machowski, J. Wide-area measurement system applied to emergency control of power system Part I Synchronous phasor measurement. *Autom. Elektroenerg.* **2005**, *2*, 26–34.
2. Zheng, X.; Wang, S.; Su, X.; Xiao, M.; Ullah, Z.; Hu, X.; Ye, C. Real-Time Dynamic Behavior Evaluation of Active Distribution Networks Leveraging Low-Cost PMUs. *Energies* **2021**, *14*, 4999. [CrossRef]
3. Binek, M. *Methods of Signal Digital Processing in Wide-Area Measurement Systems Used in Electric Power Systems*; Lodz University of Technology: Lodz, Poland, 2021. [CrossRef]
4. Binek, M.; Kanicki, A.; Rózga, P. Application of an Artificial Neural Network for Measurements of Synchrophasor Indicators in the Power System. *Energies* **2021**, *14*, 2570. [CrossRef]
5. *C37.118.1-2011*; IEEE Standard for Synchrophasor Measurements for Power Systems. Revision of IEEE Std C37.118™-2005; IEEE: Piscataway, NJ, USA, 2011; pp. 1–61. [CrossRef]
6. *C37.118.1a–2014*; IEEE Standard for Synchrophasors for Power Systems Measurements for Power Systems–Amendment 1: Modification of Selected Performance Requirements. Amendment to IEEE Std C37.118.1-2011; IEEE: Piscataway, NJ, USA, 2014; pp. 1–25. [CrossRef]

7. *60255-118-1:2018*; IEC/IEEE Standard Part 118-1: Synchrophasor Measurements for Power Systems. IEEE STD23444; International Electrotechnical Commission: Geneva, Switzerland, 2018; pp. 1–78, ISBN 978-1-5044-5361-5. [CrossRef]
8. Rakowska, A.; Grzybowski, A.; Stiller, J. Are we in danger of blackouts caused by climatic phenomena. *Energetyka* **2009**, *7*, 419–429.
9. El Mrabet, Z.; Sugunaraj, N.; Prakash Ranganathan, P.; Abhyankar, S. Random Forest Regressor-Based Approach for Detecting Fault Location and Duration in Power Systems. *Sensors* **2022**, *22*, 458. [CrossRef] [PubMed]
10. Binek, M.; Kanicki, A. Applying Artificial Neural Networks in the Identification Process of Fault Values. *Przegląd Elektrotechniczny* **2014**, *90*, 51–57. [CrossRef]
11. Binek, M.; Kanicki, A.; Korbel, P. Signal Parameters Identification Methods Used in Wide-Area Measurement Systems. In Proceedings of the 2016 Electrical Power Networks Conference—EPNet, Szklarska Poręba, Poland, 19–21 September 2016; pp. 1–5. [CrossRef]
12. Phadke, A.G.; Thorp, J.S. *Synchronized Phasor Measurements and Their Applications*; Springer: New York, NY, USA, 2008; pp. 5–195.
13. Femine, A.D.; Gallo, D.; Landi, C.; Luiso, M. The Design of a Low Cost Phasor Measurement Unit. *Energies* **2019**, *12*, 2648. [CrossRef]
14. Li, J.; Wei, W.; Zhang, S.; Li, G.; Gu, C. Conditional Maximum Likelihood of Three-Phase Phasor Estimation for µPMU in active Distribution Networks. *Energies* **2018**, *11*, 1320. [CrossRef]
15. Kang, S.-H.; Seo, W.-S.; Nam, S.-R. A Frequency Estimation Method Based on a Revised 3-Level Discrete Fourier Transform with an Estimation Delay Reduction Technique. *Energies* **2020**, *13*, 2256. [CrossRef]
16. Kušljević, M.D.; Tomić, J.J.; Poljak, P.D. On Multiple-Resonator-based Implementation of IEC/IEEE Standard P-Class Compliant PMUs. *Energies* **2021**, *14*, 198. [CrossRef]
17. Xue, H.; Cheng, Y.; Ruan, M. Enhanced Flat Window-Based Synchrophasor Measurement Algorithm for P Class PMUs. *Energies* **2019**, *12*, 4039. [CrossRef]
18. Rebizant, W.; Szafran, J. Power system frequency estimation. *IET Gener. Transm. Distrib.* **1998**, *145*, 578–582. [CrossRef]
19. Szafran, J.; Wiszniewski, A. *Measurement and Decision Algorithms of Digital Protection and Control*; WNT: Warszawa, Poland, 2001. (In Polish)
20. Thilakarathne, C.; Meegahapola, L.; Fernando, N. Static Performance Comparison of Prominent Synchrophasor Algorithms. In Proceedings of the 2017 IEEE Innovative Smart Grid Technologies—Asia, Auckland, New Zealand, 4–7 December 2017; pp. 1–6. [CrossRef]
21. Mingotti, A.; Costa, F.; Peretto, L.; Tinarelli, R. Closed-Form Expressions to Estimate the Mean and Variance of the Total Vector Error. *Energies* **2021**, *14*, 4641. [CrossRef]
22. Mostarac, P.; Malarić, R.; Mostarac, K.; Jurčević, M. Noise Reduction of Power Quality Measurements with Time-Frequency Depth Analysis. *Energies* **2019**, *12*, 1052. [CrossRef]
23. Tran, M.Q.; Zamzam, A.S.; Nguyen, P.H. Enhancement of Distribution System State Estimation Using Pruned Physics-Aware Neural Networks. *arXiv* **2021**, arXiv:2102.03893.
24. Su, H.-Y.; Liu, T.-Y. GECM-Based Voltage Stability Assessment Using Wide-Area Synchrophasors. *Energies* **2017**, *10*, 1601. [CrossRef]
25. Mirshekali, H.; Rahman Dashti, R.; Ahmad Keshavarz, A.; Shaker, H.R. Machine Learning-Based Fault Location for Smart Distribution Networks Equipped with Micro-PMU. *Sensors* **2022**, *22*, 945. [CrossRef] [PubMed]
26. Thilakarathne, C.; Meegahapola, L.; Fernando, N.P. Real-time voltage stability assessment using phasor measurement units: Influence of synchrophasor estimation algorithms. *Int. J. Electr. Power Energy Syst.* **2020**, *119*, 105933. [CrossRef]
27. Apráiz, M.; Diego, R.I.; Barros, J. An Extended Kalman Filter Approach for Accurate Instantaneous Dynamic Phasor Estimation. *Energies* **2018**, *11*, 2918. [CrossRef]
28. Lippmann, R.; Gold, B.; Malpass, M. *A Comparison of Hamming and Hopfield Neural Nets for Pattern Classification*; Lincoln Laboratory, Massachusetts Institute of Technology: Lexington, MA, USA, 1987.
29. Zamzam, A.S.; Sidiropoulos, N.D. Physics-Aware Neural Networks for Distribution System State Estimation. *IEEE Trans. Power Systems* **2020**, *35*, 4347–4356. [CrossRef]
30. Tran, M.Q.; Zamzam, A.S.; Nguyen, P.H.; Pemen, G. Multi-Area Distribution System State Estimation Using Decentralized Physics-Aware Neural Networks. *Energies* **2021**, *14*, 3025. [CrossRef]
31. Yu, Q.; Hu, Y.; Yang, Y. Identification Method for Series Arc Faults Based on Wavelet Transform and Deep Neural Network. *Energies* **2020**, *13*, 142. [CrossRef]
32. MathWorks, Inc. *Neural Network Toolbox™ User's Guide © COPYRIGHT 1984–2022*; The MathWorks, Inc. Matlab: Natick, MA, USA, 2022.
33. Lyons, R.G. *Understanding Digital Signal Processing*, 2nd ed.; Prentice Hall: Hoboken, NJ, USA, 2004.
34. Wehbe, Y.; Fan, L.; Miao, Z. Least squares based estimation of synchronous generator states and parameters with phasor measurement units. In Proceedings of the 2012 IEEE North American Power Symposium, Champaign, IL, USA, 9–11 September 2012; pp. 1–6. [CrossRef]
35. Hsieh, G.; Hung, J.C. Phase-locked loop techniques. A survey. *IEEE Trans. Ind. Electron.* **1996**, *43*, 609–615. [CrossRef]
36. Mohagheghi, S.; Venayagamoorthy, G.; Harley, R. Optimal wide area controller and state predictor for a power system. *IEEE Trans. Power Syst.* **2007**, *22*, 693–705. [CrossRef]

37. Gong, J.; Yang, W. Driver Pre-accident Behaviour Pattern Recognition Based on Dynamic Radial Basis Function Neural Network. In Proceedings of the 2011 International Conference on Transportation, Mechanical and Electrical Engineering, Changchun, China, 16–18 December 2011; pp. 328–331. [CrossRef]
38. Shah, H.; Verma, K. PMU-ANN based Approach for Real Time Voltage Stability Monitoring. In Proceedings of the 2016 IEEE 6th International Conference on Power Systems (ICPS), New Delhi, India, 4–6 March 2016; pp. 1–5. [CrossRef]

MDPI AG
Grosspeteranlage 5
4052 Basel
Switzerland
Tel.: +41 61 683 77 34

Sensors Editorial Office
E-mail: sensors@mdpi.com
www.mdpi.com/journal/sensors